# Lecture Notes in Control and Information Sciences

Edited by A. V. Balakrishnan and M. Thoma



For information about Vols. 1–21 please contact your bookseller or Springer-Verlag.

# Lecture Notes in Control and Information Sciences

Edited by M. Thoma and A. Wyner

85


# Stochastic Processes in Underwater Acoustics

Edited by C. R. Baker


Springer-Verlag
Berlin Heidelberg GmbH

**Editor**
Prof. Charles R. Baker
Department of Statistics
304 Phillips Hall
University of North Carolina
Chapel Hill, NC 27514
USA

ISBN 978-3-540-16869-0 ISBN 978-3-540-47149-3 (eBook)
DOI 10.1007/978-3-540-47149-3


Originally published by Springer-Verlag Berlin Heidelberg New York in 1986.

2161/3020-543210

## PREFACE

This volume contains seven research papers on topics in stochastic processes, with specific applications to problems in underwater acoustics. Extracts from these papers were presented in two invited sessions at the 1985 IEEE International Symposium on Information Theory, Brighton, England, June 24-28, 1985. Thanks are extended to Dr. E.C. Posner and the Symposium Program Committee for the arrangement of the sessions.

Although the focus here is on applications in underwater acoustics, it will be seen that the methods and results are also applicable to other areas. For example, impulsive noise arises in many environments, of which a specific example is acoustic noise under ice in the arctic or antarctic regions. Multipath propagation occurs in seismology and in tropospheric communications. Source localization and identification, channel modeling, tracking, signal detection, time delay estimation -- topics treated in this book -- are familiar problems in diverse areas. Models and algorithms developed in this volume can thus be used in applications other than those of underwater acoustics. With this in mind, efforts have been made to make the material readily accessible to non-specialists.

A short introductory chapter gives an overview of the volume. The remaining chapters are self-contained, with a common index.

Chapel Hill, March, 1986 — Charles R. Baker

## CONTRIBUTORS


C.R. Baker
Department of Statistics, University of North Carolina
Chapel Hill, NC 27514, USA

D. de Brucq
Laboratoire de Capteurs, Instrumentation et Systèmes
Université de Haute-Normandie, BP 67,
76130 Mont-Saint-Aignan, France

P. Duvaut
Laboratoire de Signaux et Systèmes,
Centre de recherche du CNRS et de l'ESE, associé à
l'Université de Paris-Sud, 91190 Gif-sur-Yvette, France

A.F. Gualtierotti
IDHEAP, BFSH I, Université de Lausanne
1015 Lausanne, Switzerland

A.O. Hero
Department of Electrical Engineering and Computer Science
University of Michigan, Ann Arbor, MI 48109, USA

C.S. Hwang
Department of Electrical and Computer Engineering
Oregon State University, Corvallis, OR 97331, USA

G. Jourdain
CEPHAG ENSIEG, Domaine Universitaire, BP 46
38402 Saint-Martin-d'Hères, France

R.R. Mohler
Department of Electrical and Computer Engineering
Oregon State University, Corvallis, OR 97331, USA

J M.F. Moura
CAPS, Instituto Superior Técnico
Av. Rovisco Pais, 1096 Lisbon, Portugal

M.A. Pallas
CEPHAG ENSIEG, Domaine Universitaire, BP 46
38402 Saint-Martin-d'Hères, France

B. Picinbono
Laboratoire de Signaux et Systèmes,
Centre de recherche du CNRS et de l'ESE, associé à
l'Université de Paris-Sud, 91190 Gif-sur-Yvette, France

M.J.D. Rendas
CAPS, Instituto Superior Técnico
Av. Rovisco Pais, 1096 Lisbon, Portugal

S.C. Schwartz
Department of Electrical Engineering and Computer Science
Princeton University, Princeton, NJ 08544, USA

CONTENTS

# CHAPTER 0

## INTRODUCTION

C.R. Baker

Underwater acoustics provides an environment containing some of the most challenging problems in contemporary applied stochastic processes. These problems begin with construction of realistic channel and data models and encompass development of algorithms for sonar signal detection, classification, estimation, communications, data analysis, and data enhancement. Nonstationary and nonGaussian problems are commonplace. Multivariate data processes with high dimensionality and strong correlation are the rule, rather than the exception.

Given the important practical uses of underwater acoustics, it is appropriate that much of the signal processing and modeling research is focused on near-term applications and frequently tied to specific systems. However, the complexity of the physical environment and the difficulty in constructing realistic data models emphasize the desirability of long-term research of a more basic nature.

To be useful, such research should be guided by applications and directed toward results that can eventually be incorporated into operational systems. Fortunately, the category of research that can be regarded as applied has broadened significantly during recent years. This is a consequence of the great advances in capability of computing equipment and electronics in general, enabling the use of algorithms that would have been impractical a few years ago. Further significant advances are expected. Full utilization of these capabilities will require substantial mathematical and computational research directed toward modeling, analysis, and algorithm development. Thus, there is clearly a need for long-term involvement by capable theoreticians whose research includes problems of the type described above, who are familiar with specific applications in underwater acoustics, and who are interested in the development of useful analytical results and algorithms based on their research.

Much sonar-related research involving stochastic processes is application-specific, bringing in the environment and geometry in an essential way. Examples include channel and propagation modeling which takes into account sound-speed variability and bottom-surface geometry. Other work seeks to solve problems in a general framework, motivated by applications in sonar. Examples include development of algorithms for time-delay estimation and for nonGaussian signal detection.

This range of research activity is reflected in the contents of this volume, which also reflect the diversity of research topics typical of signal processing for underwater acoustics. From an applications viewpoint, the volume proceeds through the following topics: modeling of the random pressure field; identification and modeling of multipath channels; location of a signal source; time delay estimation for a passive array; tracking of targets using noisy observations; and nonGaussian signal detection, based on likelihood ratios and on a generalized signal-to-noise ratio.

The contributors to this volume represent well-established research programs in the application of stochastic processes to problems of underwater acoustics. In addition to theoretical developments, each paper contains applications in the form of numerical results obtained from experimental data or simulations and/or in the form of explicit algorithms. The techniques used throughout are largely those of statistical communication theory and stochastic systems, combined where appropriate with considerations of ocean physics and geometry. A brief summary of each chapter follows.

Chapter 1. Identification of Causal Linear Filters and Applications in Underwater Acoustics
by D. de Brucq

Modeling of ambient noise and ocean characteristics is a fundamental problem for passive sonar, and is the topic of this chapter. Attention is focused on modeling the random pressure field associated with propagation in a deep sea environment. The approach first used is to consider the partial differential equation of the field. Results are obtained for spatial correlation at a given frequency when stationary white noise (in time and space) is the driving process. In the absence of boundary conditions (deep sea), the noise is found to be isotropic, and an expression for cross-spectral density obtained.

The model is then generalized to include a Levy process as the driving term: a random measure consisting of the sum of a stationary white Gaussian noise and Dirac measures of random amplitudes. The vector pressure process is represented as a linear filter driven by the Levy process. The problem of interest is to identify the response function of the filter and the measure defined by the Levy process, under the assumption that the random amplitudes of the Dirac measures are i.i.d. spherically-invariant random variables. This is solved by applying the method of moments. The paper concludes with a summary of numerical results obtained in an analysis of experimental data.

Chapter 2. Multiple Time Delay Estimation in Underwater Acoustic Propagation

by G. Jourdain and M.A. Pallas

Another basic modeling problem in underwater acoustics is that of characterizing the channel between a transmitter and a receiver. This problem is analyzed here using the approach of linear systems theory; the channel is assumed to be a deterministic linear filter acting in the presence of additive Gaussian noise. The criteria used are minimum mean square error and maximum likelihood.

These general considerations are applied to active identification of a multipath channel using PSK signals modulated by maximal length binary sequences. The channel is represented as the sum of several paths, each characterized by a time delay, phase shift and amplitude attentuation. The number of paths is unknown. Optimum estimators for these quantities are developed, and bounds on their variance are determined. Joint estimation of time delay and phase shift is emphasized, with amplitudes treated as essentially known. Experimental results are given for an experiment which produces three main paths, and comparisons made with ray tracing. Results obtained for the time delays of these paths, based on 450 realizations, are found to have a sample variance for the estimated time delays which compares well with the Cramer-Rao lower bound. Estimation of phase shift is seen to be a more difficult problem.

Chapter 3. Optimal Filtering in the Presence of Multipath

by J.M.F. Moura and M.J.D. Rendas

The problem treated in this chapter is that of locating the position of a signal source which is observed by passive sonar through a multipath channel. The authors begin by considering the problem of modeling the channel. Several situations are analyzed, depending on assumptions such as sound speed and environment (reflecting boundaries). A linear filter model is used for the ocean, consisting of the sum of several paths, each introducing time delay and amplitude attenuation. An algorithm for determining ray structure is obtained, giving estimates of the number of paths and of the attenuation and time delay along each path. This algorithm is a function of the sound speed profile, source and receiver locations, and the medium geometry.

The problem of determining source location is then analyzed. Maximum likelihood estimation is considered for a Gaussian source in the presence of additive Gaussian noise. The general situation is first considered. Attention is then restricted to two different hypotheses on the signal: when the signal is contained in a finite-dimensional subspace with known basis functions, and when it is represented as the output of a finite-dimensional linear system driven by white noise. Maximum-likelihood estimation is applied for the first problem, and Kalman-Bucy filtering techniques are applied to the second. Receiver structure is analyzed and representations presented for each of these signal classes.

Chapter 4. Level Crossing Representations, Poisson Asymptotics and Applications to Passive Arrays

by A.O. Hero and S.C. Schwartz

Level crossing problems arise naturally in sonar. One obvious example is the number of false alarms in a given time period (false alarm rate), resulting from a test statistic exceeding a threshhold. However, less apparent level crossing problems also occur, one of which is treated by the authors of this chapter.

The authors begin by obtaining a new result for the probability of obtaining one or more upcrossings of zero in a finite time interval by a path-continuous nonstationary stochastic process. They also obtain an asymptotic result for a normalized counting process to converge (in distribution) to a nonstationary Poisson process. These results are then applied to the problem of estimating the delay in arrival time between two hydrophones of a signal field wavefront for a Gaussian signal in additive Gaussian noise. The time delay at which the sample cross-correlation peaks is the estimate. This estimate is asymptotically equivalent to the maximum likelihood estimate as the time-bandwidth product increases. The authors treat the case of large errors; in this situation, ambiguous peaks occur. A global variance approximation for the estimator is obtained. Results on estimation errors are obtained, and related to the Cramer-Rao and Zakai-Ziv bounds. Numerical results for specific examples are presented.

Chapter 5. Nonlinear Data Observability and NonGaussian Information Structures

by R.R. Mohler and C.S. Hwang

The problem considered is that of reconstructing a process x(t) from measured data y(t), where y has a functional relationship to x. In underwater acoustics, noise is usually present, and the problem is substantially more difficult.

Deterministic nonlinear observability is first considered here, and applied to two sonar problems: bearing-only-target tracking, and linear array tracking. Stochastic observability is then analyzed in terms of the (Shannon) mutual information between an a priori unobserved dynamic process and an a posteriori observed measurement process. The results are applied to a detailed analysis of the two sonar tracking problems, taking into account sensor-target geometry and target maneuvering. This provides relations between observability of range and target velocity and such quantities as number and type of sensors. These results are illustrated by extensive simulations, which show the importance of Doppler. Relations between filtering error and information are also obtained. Advantages of the Shannon information approach over the Fisher information approach are summarized.

Chapter 6. Likelihood Ratios and Signal Detection for NonGaussian Processes.

by C.R. Baker and A.F. Gualtierotti.

The emphasis is on development of likelihood ratios and detection algorithms for problems involving nonGaussian data. The first problem considered is that of detecting a nonGaussian signal in Gaussian noise. This frequently arises in active sonar; it could also be important for passive sonar. General results are presented on non-singular detection and likelihood ratio. A recursive discrete-time detection algorithm is obtained and is shown to be a likelihood ratio detector when the signal-plus-noise is Gaussian.

The second major problem considered is that of detecting a signal in spherically-invariant noise (SIN). This is a model which has been proposed for some impulsive-plus-Gaussian environments, and is closely linked to detection problems encountered in some active sonar applications. General results on nonsingular detection and likelihood ratio are first obtained. For detection of a known signal, the behavior of the discrete-time likelihood ratio is analyzed as the sample size increases. Constant-false-alarm-probability detectors are given, and an example based on sonar data illustrates the potential loss due to using a Gaussian model when the noise is actually nonGaussian SIN.

Chapter 7. Detection and Contrast

by B. Picinbono and P. Duvaut

Optimum signal detection, under various measures of optimality, involves evaluation of the likelihood ratio. However, implementation of the likelihood ratio requires information about the data which is frequently not available in underwater acoustics applications. Alternative sub-optimum detection procedures and criteria, such as contrast, are thus of interest. The contrast of a filter (with data vector x) S(x) is defined as $[E_1(S) - E_0(S)]^2/V_\Pi(S)$ where $E_1(\cdot)$ denotes expectation when signal is present, $E_0(\cdot)$ expectation for noise only, and $V_\Pi(\cdot)$ is the variance for a prior distribution $\Pi$, with expectation $(1-\pi)E_0(\cdot) + \pi E_1(\cdot)$.

The authors first discuss several properties of the contrast, including the relation to asymptotic relative efficiency, invariance, and relation to singular detection. Optimal properties of the contrast are then considered. A major result is the following. A normalized filter R is defined in terms of the likelihood ratio and the prior distribution. Minimum distance from R to any normalized family of filters F is shown (under Hilbert subspace conditions) to be attained by the element of F having maximum contrast. Following a discussion of the consequences of this result, the effect of monotone transformations and quantization on contrast and detection is considered and numerical results are presented for a specific monotone transformation.

As a collection of research papers, this volume is not intended to be an introduction to stochastic processes in underwater acoustics. For an introduction to the area, and a broader perspective, the reader is referred to the books listed below.

L.M. Brekhovskikh and Yu. Lysanov, Fundamentals of Ocean Acoustics Springer-Verlag, Berlin (1982)

S.M. Flatte (ed.), R. Dashen, W.H. Munk, K.M. Watson, and F. Zachariasen, Sound Transmission through a Fluctuating Ocean, Cambridge University Press, Cambridge (1979).

V.V. Ol'shevskii, Statistical Methods in Sonar, Consultants Bureau, New York (1978).

R.J. Urick, Principles of Underwater Sound, McGraw-Hill, New York (1975).

L.J. Ziomek, Underwater Acoustics: A Linear Systems Theory Approach, Springer-Verlag, Berlin (1982).

CHAPTER 1

# IDENTIFICATION OF CAUSAL LINEAR FILTERS AND APPLICATIONS IN UNDERWATER ACOUSTICS

Denis de Brucq

## 1. INTRODUCTION

The aim of this chapter is to model the random phenomena of underwater acoustics propagation in order to arrive at a better understanding of the observations measured by sensors.

It is known that information can be transmitted at low frequencies over hundreds of kilometers in spite of random attentuations of the acoustic wave. The phase velocity c depends on the frequency $\nu$ and also on temperature, pressure and salinity, so is random as a function of random variables.

Geometric attenuations are usually greater than the dissipations for acoustic propagation. However we will show that the new term $b \frac{\partial}{\partial t} \Delta p$ in the equation of propagation (2-15) will fully explain the correlation $\gamma$ of the random pressure field p at a given frequency $\nu$. The theory presented in §2 is not specific to this problem and can be applied in other areas of random physics. This theory explains the second order properties of acoustic noise in the deep sea.

For flickering sources localized in a given position, we use as the driving term of the propagation equation a general Levy process L including a Gaussian process W and thus deduce the statistical law of the observations.

A notation index is included to clarify formulas.

Independence of noise and signal sources is assumed. All processes are centered, second order and normalized to have zero mean and unit variance.

The observation Z is the pressure at the sensors. The number of sensors varies from ten to several thousands and each coordinate of Z is linked to one localized sensor. As the observation is not continuous in time, we have at our disposal a vector process

$$Z = (\Omega, A, P, (Z(t))_{\mathbb{R}}, \quad (\mathbb{R}, R)^{\otimes q})$$

where $(\Omega, A, P)$ is a probability space, $(\mathbb{R}, R)$ is a classical measurable real space with Borelian sets $R$ and q is the number of sensors.

For such a second order centered process Z, the correlation function $\Gamma$ and the spectral density function $\rho$ are introduced; here the process is assumed to be stationary for every real t, $\tau$,

$$\Gamma(\tau) = E(Z(t)\ Z(t-\tau)^*) = \int e^{-i\nu\tau}\ \rho(\nu)\ d\nu.$$

The $q \times q$ matrix spectral density function $\rho$ is estimated by sampling. To achieve a good approximation, the frequency band [0,B] is defined, implying the sampling time $\delta T \triangleq 1/2B$. The covariance is estimated by averaging of N sample observations for n values: $\tau = 0, \Delta T, \ldots, (n-1)\ \Delta T$. The Fourier transform gives an estimate of $\rho$ for $\nu = 0, B/n, \ldots, (n-1)\ B/n$.

This procedure thus requires estimates of $q(q+1)n/2$ coefficients. The data gathering and computational requirements for this are formidable.

Another means of estimating $\rho$ is to suppose Z to be Auto Regressive with Moving Average (ARMA), i.e., satisfying the linear equation:

$$Z(t) + a(1)\ Z(t-1) + \ldots + a(p)\ Z(t-p) = b(0)\ e(t) + \ldots + b(r)\ e(t-r)$$

where $a(1), \ldots, a(p)$ are p matrices $q \times q$

$b(0), \ldots, b(r)$ are r matrices $q \times s$

$(e(t), t \in Z)$ is a white noise in $(\mathbb{R}, R)^{\otimes s}$.

Here the time t is integer with unit $\Delta T$.

Noting the transfer function of the ARMA filter,

$$G(z) \triangleq (I+a(1)\ z^{-1}+\ldots+a(p)\,z^{-p})^{-1}(b(0) + b(1)\ z^{-1}+\ldots+b(r)\ z^{-r})$$

the spectral density function is given by

$$\rho(\nu) = \frac{1}{2\pi}\ G(e^{-i\nu})\ G(e^{-i\nu})^*.$$

The coefficients to be identified are then the entries of the matrices $(a(i))$, $(b(j))$; if the number of such coefficients is considered when q = 1,000, p = r = s = 10, it becomes obvious why the ARMA model cannot be used directly in underwater acoustics. It is essential to simplify the model to those parameters which are physically meaningful.

The detection in a given direction of the number K of sources is another very important problem which must be solved before the sources are identified. The q-dimensional observation Z is the sum of independent processes, signals and noise:

$$Z = S(1) + \dots + S(K) + N.$$

Here the noise N is the sum of the random pressure fluctuation in the sea and the electronic noise from the amplification system.

The classical estimation of the number K of sources assumes that they are uncorrelated and that the q components of the noise have the same spectral density function.

Each signal $S(\ell)$ $\ell = 1, 2, \dots, K$ comes from a random source $X(\ell)$ through a random medium. The random source $X(\ell)$ is a second order stationary scalar process and a vector impulse response describes the geometric properties of the propagation from the source to the sensors. Thus, for all real t,

$$S(\ell,t) = \int_{-\infty}^{t} G(\ell, t-\tau)\, X(\ell,\tau)\, d\tau.$$

The Fourier transform then gives

$$Z(\nu) = \sum_{\ell=1}^{K} G(\ell,\nu)\, X(\ell,\nu) + N(\nu).$$

The covariance matrix at a given frequency $\nu$ is readily computed

$$\Gamma(\nu) \triangleq E(Z(\nu)\, Z(\nu)^*)$$

$$= \sum_{\ell=1}^{K} G(\ell,\nu)\, G(\ell,\nu)^*\, \sigma^2(\ell,\nu) + I\, \sigma^2(N,\nu)$$

where I is the $q \times q$ identity matrix, and

$$\sigma^2(\ell,\nu) \triangleq E(|X(\ell,\nu)|^2)$$

$$\sigma^2(N,\nu)\, I \triangleq E(N(\nu)\, N(\nu)^*).$$

Note that each sensor is perturbed by the same noise. In practice, the Fourier transform $Z(\nu)$ is obtained by sampling and the Fast Fourier Transform; the matrix $\Gamma$ is obtained from the product $Z(\nu)\, Z(\nu)^*$ followed by averaging. Because of the presence of the diagonal matrix $I\sigma^2(N,\nu)$, the rank of $\Gamma$ is q.

The eigenvalues are ordered by decreasing eigenvalues and the sources with energy $\sigma^2(\ell,\nu)$ greater than the noise energy $\sigma^2(N,\nu)$ are detected.

In Section 2, we model the noise for acoustic propagation in the deep sea. We introduce the equation

$$\text{(2-12)} \qquad \Delta p - \frac{1}{c^2}\frac{\partial^2}{\partial t^2} p + \frac{b}{c^2}\frac{\partial}{\partial t}\Delta p = dW$$

for the random pressure p. The second member dW is a random distribution white noise in spatio-temporal space: for $\phi, \psi \in L^2(\mathbb{R}^4, R^4, dx\, dy\, dz\, dt)$,

$$E[\int\phi\, dW \int\psi\, dW] = \int \phi\, \psi\, dx\, dy\, dz\, dt.$$

If $\phi$ and $\psi$ have disjoint supports, the correlation of $W(\phi) \triangleq \int \phi\, dW$ and $W(\psi) \triangleq \int \psi\, dW$ is zero. Experimental curves [9] show that the equation (2-12) represents a whitening of the noise p in the deep sea.

With no boundary conditions, we deduce that the noise is isotropic. The interspectral density $\gamma(D,\nu)$ depends only on the distance D of the two points and

$$\text{(2-28)} \qquad \gamma(D,\nu) = C\, \frac{e^{-\alpha D}}{\alpha}\, \frac{\sin kD}{kD}$$

with the wavenumber $k(\nu)$, the attentuation $\alpha(\nu)$ at frequency $\nu$, and a normalization coefficient, C.

This result applies for any equation

$$P\,(\frac{\partial}{\partial t}) + Q\,(\frac{\partial}{\partial t})\,\Delta p = dW$$

with P and Q polynomials and for any white process dW, not necessarily Gaussian.

For D = 0 the relation reduces to

$$\rho(\nu) \triangleq \gamma(0,\nu) = C/\alpha(\nu) \quad \text{or} \quad \log \rho(\nu) = -\log \alpha(\nu) + \log C,$$

showing a simple relation between the spectral density function $\rho(\nu)$ and the attentuation coefficient $\alpha(\nu)$.

Up to now we have considered only second order, stationary properties of the pressure field p. Instead of having stationary white noise dW as the driving term, localized sources, possibly nonstationary in time, can be be considered. In Section 3, Levy processes L are introduced. In sonar theory the threshold is continuously adjusted. An estimation of the noise energy is constantly carried out to optimize the threshold. This indicates that the process Z is not Gaussian and measured noise statistics confirm that result.

If we solve equation (2-12) with a Levy process driving term, we obtain an explicit solution with a kernel G which is homogeneous in space and in time:

$$p(x, y, z, t) = \int G(x-\xi, y-\eta, z-\zeta, t-\tau)\, dL(\xi, \eta, \zeta, \tau).$$

If the pressure p is observed in q points

$(x(1), y(1), z(1)) \ldots (x(q), y(q), z(q))$,

then we write

$$Z(\ell,t) = p(x(\ell), y(\ell), z(\ell), t) \quad \ell = 1, 2, \ldots, q$$

$$Z(t) = (Z(1,t), \ldots, Z(q,t))^* \quad \text{and}$$

$$Z(t) = \int G(\xi, \eta, \zeta, t-\tau)\, dL(\xi, \eta, \zeta, \tau)$$

with a vector G.

For localized sources, spatial integration is replaced by a finite sum:

$$Z(t) = \sum_{\ell=1}^{K} \int_{-\infty}^{t} G(\ell, t-s)\, dL(\ell, s) \tag{3-23}$$

with obvious notations: $L(\ell,\cdot)$ is the Levy process and $G(\ell,\cdot)$ is the impulse response of the $\ell^{th}$ source.

A Levy process L is defined by the second characteristic function which depends upon a measure $\lambda$ of the jumps. A Levy process L is a random measure:

$$dL(s) = dW(s) + \sum_{i \in Z} A(i)\, \delta(s(i), ds).$$

The Dirac measures $\delta(\cdot, ds)$ are randomly distributed. The random sequence $(s(i), i \in Z)$ constitutes a Poisson point process. The $A(i)$, $i \in Z$, are random, identically distributed, independent variables and independent of the other processes. A Levy process can be white.

In Section 4, spherically invariant laws for A are considered. These are infinite convex mixtures of centered Gaussian laws. In addition to the Levy case, direct models of sea pressure using these laws are indicated. The theoretical methods for achieving identification with the above descriptions are then described.

From the various techniques available for estimating a measure $\nu$, we select the moments technique. Convex sums of Dirac measures are written:

(S-1) $d\nu(b) \triangleq \nu(1)\ \delta(b(1),\ db) + \ldots + \nu(n)\ \delta(b(n),\ db)$

so that the first 2n moments are the same as for the measure $\nu$. Various improvements and applications are given in §5.

Finally, we recall the auto-regressive technique for identifying a causal linear filter from the covariance of the output [19]. We consider sea pressure observations and a Pascal program created to obtain identification parameters. Lack of space has meant that we are able to give only a brief outline of the practical difficulties to be solved.

## 2. SPATIAL CORRELATION AT A GIVEN FREQUENCY

The idea is to use general equations of fluid dynamics [15] to obtain the pressure noise correlation in underwater acoustic. The first approach would be to start from the equation

$$\Delta p - \frac{1}{c^2}\frac{\partial^2}{\partial t^2} p = 0 \tag{1}$$

and to suppose that this equation whitens the pressure noise, i.e.,:

$$\Delta p - \frac{1}{c^2}\frac{\partial^2}{\partial t^2} p = dW \tag{2}$$

where dW is a white random distribution.

The meaning of this last equation will be clearer if the simpler case is considered:

$$d\dot{y} + b\dot{y}\ dt + cy\ dt = dW \tag{3}$$

where b, c are constants and where y is a real process with correlation to be determined from the usual white noise process dW.

The solution space for

$$\ddot{y} + b\dot{y} + cy = 0 \tag{4}$$

is spanned by

$$\begin{cases} y(1,t) \triangleq \exp\ (-\ \alpha t)\ \exp\ (2\ i\pi\nu t) \\ y(2,t) \triangleq \exp\ (-\ \alpha t)\ \exp\ -\ (2i\pi\nu t) \end{cases} \tag{5}$$

where $r(1) \triangleq -\alpha + 2i\pi\nu$ and $r(2) \triangleq -\alpha - 2i\pi\nu$ are the two solutions to

$$r^2 + br + c = 0. \tag{6}$$

We assume that $b^2 - 4c < 0$ and $b < 0$ in order to have the solutions with negative real part.

By classical argument [16, p. 2]] the equation

$$\ddot{y} + b\dot{y} + cy = \delta \tag{7}$$

has the solution

$$y(t) = Y(t)\ \exp - (\alpha t)\ \sin(2\pi\nu t)/2\pi\nu$$
$$= Y(t)\ (y(1,t) - y(2,t))/4i\pi\nu,$$

where $Y(t) \overset{\Delta}{=} 1_{[0,\infty[}(t)$ is the Heaviside Function.

The causal Green's function

$$G(t-\tau) \overset{\Delta}{=} Y(t-\tau)\ \exp - \alpha(t-\tau)\ \sin 2\pi\nu\ (t-\tau)/2\pi\nu \tag{8}$$

is used to solve any equation $\ddot{y} + b\dot{y} + cy = F$ with the driving term F. The solution is $y(t) = \int_{-\infty}^{t} G(t-\tau)\ F(\tau)\ d\tau$ and, if the driving term is the white noise dW, we get

$$Y(t) = \int_{-\infty}^{t} G(t-\tau)\ dW(\tau). \tag{9}$$

The second order, centered stationary process Y(t) has the correlation function

$$\Gamma(\tau) \overset{\Delta}{=} E(Y(t)\ Y(t-\tau)) \tag{10}$$
$$= \frac{1}{4}\frac{e^{-\alpha\tau}}{\alpha}\left\{\frac{\cos(2\pi\nu\tau)}{\alpha^2 + (2\pi\nu)^2} + \frac{\alpha}{\alpha^2 + (2\pi\nu)^2}\frac{\sin(2\pi\nu\tau)}{2\pi\nu}\right\}.$$

We observe that the expression of $\Gamma$ depends directly on the attenuation coefficient $\alpha$ and the frequency coefficient $\nu$ of the solutions (5) to equation (4).

If $\tau = 0$, we have

$$E(Y(t)^2) = \Gamma(0) = (\alpha^2 + (2\pi\nu)^2)^{-1}(4\alpha)^{-1}. \tag{11}$$

We see that $\alpha$ must be strictly positive; otherwise equation (3) has no finite solution. Thus, the dissipation term b has to be strictly positive.

The reason why equation (2) is not satisfactory is now clear and we introduce [4, p. 24] the partial differential equation:

$$\Delta p - \frac{1}{c^2}\frac{\partial^2}{\partial t^2} p + \frac{b}{c^2}\frac{\partial}{\partial t}\Delta p = dW. \tag{12}$$

We need a dissipation term but its exact form, actually $\frac{b}{c^2}\frac{\partial}{\partial t}\Delta p$, is not critical.

The spherical solutions

$$p(D,t) \triangleq \exp -(\alpha D) \exp i(2\pi\nu t - kD)/D \tag{13}$$

where

$$D^2 \triangleq x^2 + y^2 + z^2 \tag{14}$$

of the equation

$$\Delta p - \frac{1}{c^2}\frac{\partial^2}{\partial t^2} p + \frac{b}{c^2}\frac{\partial}{\partial t}\Delta p = 0 \tag{15}$$

will give, as previously, the parameters of the correlation function $\gamma(D,\tau)$ at a given frequency $\nu \triangleq \tau/2\pi$ for the solution process p to (12).

Here we use the Fourier transform to obtain

$$p(\xi, \eta, \zeta, \tau) \triangleq \int \exp i(\xi x + \eta y + \zeta z + \tau t)\ p(x,y,z,t) dx\, dy\, dz\, dt \tag{16}$$

instead of the unknown Green's function for equation (12).

A relation between

$$\begin{cases} \tau \triangleq 2\pi\nu \text{ and} \\ \delta \triangleq -k + i\alpha \end{cases} \tag{17}$$

is needed for p(D,t) to be the solution to (15).

Putting (13) in (15), we find

$$((i\delta)^2 - (i\tau)^2/c^2 + b(i\tau)\ (i\delta)^2/c^2)\ p(D,t) = 0.$$

We denote

$$A(\delta,\tau) \triangleq \tau^2/c^2 - \delta^2(1 + ib\tau/c^2) \tag{18}$$

and if

$$A(\delta,\tau) = 0 \text{ then } p(D,t) \text{ is the solution to (15).} \tag{19}$$

Because of the complex term $-ib/c^2$, the variable $\delta$ is a complex function of $\tau$. Let $\delta(1) \triangleq -k + i\alpha$ and $\delta(2) \triangleq -\delta(1)$ be the two solutions to (18) with $A(\delta,\tau) = 0$. Then

$$A(\delta,\tau) = -(1 + ib\tau/c^2)\ (\delta^2 - (-k + i\alpha)^2). \tag{20}$$

Thus

$$(-k + i\alpha)^2 = \tau^2/(1 + ib\tau/c^2)c^2$$

$$\cong \tau^2 \, (1 - ib\tau/c^2)/c^2.$$

A first order approximation with respect to b gives

$$-k + i\alpha \cong -\tau(1 - ib\tau/2c^2)/c \text{ as } k > 0.$$

Thus, $k \cong \tau/c$ and $\alpha \cong b\tau^2/2c^3$.

Now we compute the correlation properties of the process p solving (12): For every real X, Y, Z, and T,

$$\Gamma(X, Y, Z, T) \triangleq \tag{21}$$

$$= E(p(x, y, z, t) \; p(x-X, y-Y, z-Z, t-T)$$

$$= \int \exp i(\xi X + \eta Y + \zeta Z + \tau T) \; \rho(\xi, \eta, \zeta, \tau) \; d\xi \, d\eta \, d\zeta \, d\tau$$

where $\rho$ is a spatial temporal spectral density function.

We denote

$$\delta^2 \triangleq \xi^2 + \eta^2 + \zeta^2 \text{ and} \tag{22}$$

$$D \triangleq \Delta - \frac{1}{c^2} \frac{\partial^2}{\partial t^2} + \frac{b}{c^2} \frac{\partial}{\partial t} \Delta \,. \tag{23}$$

For every x, y, z, t $\in \mathbb{R}$ and for every tempered distribution T (usually written $T \in S'$), we obtain by direct verification

$$DF(T)\,(x, y, z, t) = F(AT)\,(x, y, z, t).$$

For example,

$$\frac{\partial}{\partial x} \int \exp\,(i\xi x)\, T d\xi = \int \exp\,(i\xi x)\; i\xi T \, d\xi.$$

Using a general mathematical result [6, p. 216],

$$\rho(\xi, \eta, \zeta, \tau) = \sigma^2/|A(\delta,\tau)|^2 \tag{24}$$

where $\sigma^2$ is the constant spatial temporal spectral density function of the white noise dW. Heuristically

$$\sigma^2 \, d\xi \, d\eta \, d\zeta \, d\tau = E(|FdW|^2) \,.$$

The initial operator D is invariant under space rotation, similarly s and $\rho$ are also invariant under space rotation, and depend on $\delta^2 \triangleq \xi^2 + \eta^2 + \zeta^2$; thus $\Gamma$, the Fourier transform of $\rho$ (21), is invariant under space rotation. Using the Fubini theorem,

$$\Gamma(D,T) = \Gamma(X, Y, Z, T) \tag{25}$$
$$= \int \exp (i\tau T)\, d\tau \int \exp i(\xi X + \eta Y + \zeta Z)\ \rho(\xi, \eta, \zeta, \tau)\ d\xi\, d\eta\, d\zeta$$
$$\triangleq \int \exp (i\tau T)\ \gamma(D,\tau)\ d\tau$$

where $\gamma(D,\tau)$ is the covariance of the observations at two points distant from D, filtered at the frequency $\nu = \tau/2\pi$.

We must now give the analytical expression of $\gamma(D,T)$. We use spherical coordinates [8, p. 39] to get

$$\gamma(D,\tau) \triangleq \int \exp i(\xi X + \eta Y + \zeta Z)\ \sigma^2/|A(\delta,\tau)|^2\ d\xi\, d\eta\, d\zeta \tag{26}$$
$$= 4\pi\sigma^2 \int_0^\infty (\sin (\delta D)/\delta D . |A(\delta,\tau)|^2)\ \delta^2\, d\delta$$
$$= 4\pi\sigma^2 \int_{-\infty}^{+\infty} (\exp (i\delta D)/2i\delta D . |A(\delta,\tau)|^2)\ \delta^2\, d\delta.$$

Calculus of residues gives the integral. The poles come from the denominator, cf. (20):

$$|A(\delta,\tau)|^2 = (1 + (b^2\tau^2/c^4)) \times$$
$$(\delta + k - i\alpha)\ (\delta - k + i\alpha)\ (\delta + k + i\alpha)\ (\delta - k - i\alpha).$$

The path integration is shown on the drawing

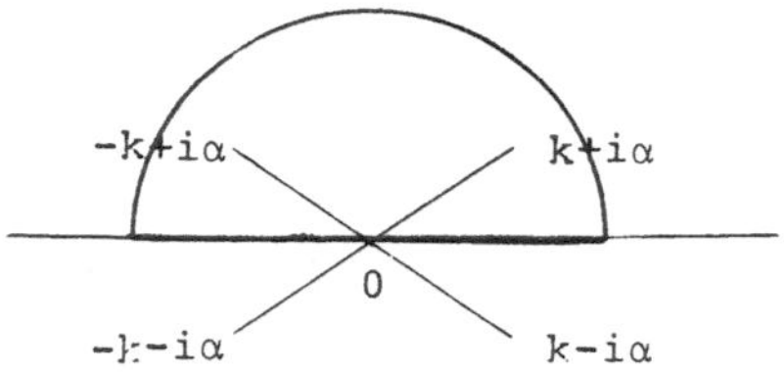


We denote

$$C/D \triangleq (4\pi\sigma^2/2iD)\ (2i\pi/(1 + (b^2\tau^2/c^4))) \cong 4\pi^2\ \sigma^2/D. \tag{27}$$

Then

$$\gamma(D,T) = (C/D)\ (\mathrm{Res}\ (-k + i\alpha) + \mathrm{Res}\ (k + i\alpha)) \tag{28}$$
$$= C\ (\exp\ (-\ \alpha D)/\alpha)\ (\sin\ (kD)/kD).$$

Usually the white noise dW is Gaussian. In fact, only the spectral property of the white noise is needed for the above theory and the statistical law of dW does not matter. In the next section, we generalize to non-Gaussian white noise and even to nonstationary processes for which dL will denote the driving term of (12).

## 3. LEVY PROCESS DRIVING TERM

Suppose the observation Z is a function of the source process through the expression

$$Z(t) = \int_{-\infty}^{t} G(t-s)\, dL(s). \tag{1}$$

Until the precise definition of a Levy process is given, one can consider L as a Wiener process W.

In fact, the vector process Z, which gives the pressure at q points, is not directly observable. The sensors with their amplification systems and the usual linear filters (bandwidth or narrowband filter, sampling filter, Fast Fourier Transform, etc.) change the initial vector process Z into a random variable. A function f is used to denote this last operation and using

$\langle f, G\rangle\ (s) \triangleq \int f(t)\ G(t-s)\ dt$, we have

$$\begin{aligned} \langle f, Z\rangle &= \int f(t)\ Z(t)\ dt \\ &= \int \langle f, G\rangle\ (s)\ dL(s). \end{aligned} \tag{2}$$

When dL is the Wiener measure dW, we can easily compute the characteristic functions of the centered Gaussian variable $\langle f, Z\rangle$:

$$\begin{aligned} E\ (\exp i \langle f, Z\rangle) &= \exp - (E\ (\langle f, Z\rangle^2)/2) \\ &= \exp - (\textstyle\int \langle f, G\rangle\ (s)^2\ ds/2). \end{aligned}$$

To obtain the classical characteristic $\phi$ we only change f into u f with u in $\mathbb{R}$. Then

$$\begin{aligned} \phi(u) &= E(\exp iu \langle f, Z\rangle) \\ &= \exp - (u^2 \textstyle\int \langle f, G\rangle^2 (s) ds/2) \end{aligned} \tag{3}$$

and the second characteristic function $\psi$ is

$$\psi(u) \triangleq \log \phi(u) = - u^2 \int \langle f, G\rangle^2 (s) ds/2. \tag{4}$$

For a centered Gaussian variable $\langle Z, f\rangle$, the moments of which satisfy the equalities: $\forall\ n \in \mathbb{N}$

$$E(\langle f, Z\rangle^{2n+1}) = 0 \tag{5}$$

$$E(\langle f, Z\rangle^{2n}) = (E\langle f, Z\rangle^{2})^{n} \cdot (2n!)/2^{n}\, n! \, . \tag{6}$$

Experimental data [5, p. 1216] show that the Gaussian law does not fit with every situation. We thus propose a Levy process as the driving term. It is a distribution-valued process [10, p. 36]

$$dL(s) \triangleq dW(s) + \sum_{i \in Z} A(i)\ \delta(s(i), ds). \tag{7}$$

The Wiener measure dW and the random measure $\sum_{i \in Z} A(i)\delta(s(i), ds)$ are independent. The sequence $(s(i);\ i \in Z)$ constitutes a Poisson point process on $\mathbb{R}$. The sequence $(A(i);\ i \in Z)$ of independent identically distributed random variables is independent of the above two processes.

Definition (7) makes clear equation (1):

$$Z(t) = \int_{-\infty}^{t} G(t-s)\ dW(s) + \sum_{s(i) \leq t} A(i)\ G(t-s(i)), \tag{8}$$

although this is not the most general Levy Process.

The general centered second order Levy Process is defined with the help of two positive measures $\mu$, $\nu$.

For every function g satisfying

$$\int_{\mathbb{R}} g(s)^2\ d\nu(s) + \int_{\mathbb{R}\times\mathbb{R}} g(s)^2\ a^2\ d\mu(a,s) < \infty, \tag{9}$$

we consider

$$L(g) \triangleq \int_{\mathbb{R}} g(s)\ dL(s) \tag{10}$$

a centered random variable with second characteristic function

$$\psi(L(g)) \triangleq - \int_{\mathbb{R}} g(s)^2\ d\nu(s) \tag{11}$$

$$+ \int_{\mathbb{R}\times\mathbb{R}} (\exp\ (iag(s)) - 1 - iag(s))\ d\mu(a,s).$$

The existence of a probability space $(\Omega, \mathcal{A}, P)$, such as the linear process $g \to L(g)$ with a probability law described by (11), is a mathematical problem [7, p. H-III-8] not studied here. The measure $\nu$ on $(\mathbb{R}, \mathcal{R})$ gives the Gaussian part of the process which is independent of the point process given by the measure $\mu$ on $(\mathbb{R}\times\mathbb{R}, \mathcal{R}\otimes\mathcal{R})$.

If $\nu = 0$ and

$$\mu(da, ds) \triangleq (\mu(1)\ \delta(a(1), da) + \ldots + \mu(n)\ \delta(a(n), da)) * \alpha\, ds,$$

then

$$\psi(L(g)) = \alpha \sum_{\ell=1}^{n} \int \mu(\ell)\ (\exp ia(\ell)\ g(s) - 1 - ia(\ell)\ g(s))\, ds,$$

and for $g(s) \triangleq u\, 1_{[a,b]}(s)$ and a, b, u in $\mathbb{R}$,

$$\psi(u) = \log E\ (\exp iuL\ (1_{[a,b]}))$$

$$= \sum_{\ell=1}^{n} \mu(\ell)\ \alpha(b-a)\ (\exp iua(\ell) - 1 - iua(\ell)).$$

For an interval [a,b] we obtain the second characteristic function of n independent variables, each with Poisson law, and mixed with probabilities $\mu_1, \ldots, \mu_n$.

Each Poisson law with parameter $\alpha(b-a)$ has jumps equal to $a(\ell)$, and is centered.

This law is also the law of the product AP where P is a centered Poisson variable with the parameter $\alpha(b-a)$ and A an independent simple variable equal to $a(\ell)$ with probability $\mu(\ell)$ for $\ell = 1, 2, \ldots, n$. From the two measures $\mu$ and $\nu$, we define

$$d\lambda(a,s) \triangleq \delta(da) \otimes d\nu(s) + a^2\, d\mu(a,s). \tag{12}$$

Then

$$\psi(L(g)) = \int (\exp\ (iag(s)) - 1 - iag(s))/a^2\ d\lambda(a,s). \tag{13}$$

We see that the Gaussian part is the limit of the jump process as the jump sizes tend to zero.

Coming back to (2), we use (13) with $g(s) \triangleq \langle f, G\rangle(s)$ to define the law of the observation process Z:

$$\psi(\langle f, Z\rangle) = \int (\exp\ (ia\langle f, G\rangle) - 1 - ia\langle f, G\rangle)/a^2\ d\lambda(a,s) \tag{14}$$

with the linear process $f \to \langle f, Z\rangle$. Classical arguments establish the law of $Z(i(1), t(1)), \ldots, Z(i(n), t(n))$, coordinate $i(1)$ of variable $Z(t(1))$, and so on. There is no theoretical difference between space and time and to simplify notation we suppose Z to be scalar.

For every $u \triangleq (u(1), \ldots, u(n) \in \mathbb{R}^n$ and for every $(t(1), \ldots, t(n)) \in \mathbb{R}^n$ let

$$\langle u, Z\rangle \triangleq u(1)\ Z(t(1)) + \ldots + u(n)\ Z(t(n)). \tag{15}$$

Generalizing from function f to distribution

$$f \triangleq u(1)\ \delta(t(1),\cdot) + \ldots + u(n)\ \delta(t(n),\cdot),$$

we easily obtain:

$$\langle f, G\rangle = u(1)\ G(t(1)-s) + \ldots + u(n)\ G(t(n)-s) \tag{16}$$

$$\triangleq \langle u, G\rangle\ (s)$$

and since (2) applies,

$$\psi(u) = \log E\ (\exp i \langle u, Z\rangle) \tag{17}$$

$$= \int (\exp (ia \langle u, G\rangle - 1 - ia \langle u, G\rangle))/a^2\ d\lambda(a,s).$$

The measure $\lambda$ and the impulse response G of a causal linear filter determine the law of the observation process. The identification problem consists in estimating the measure $\lambda$ and the response G from experimental data. This can be approached in several ways; in spite of some practical difficulties we select the method of moments.

We thus assume the existence of moments of all orders and develop the characteristic functions with respect to u:

$$\phi(u) = E\ (\exp i \langle u, Z\rangle) \tag{18}$$

$$= \sum_{p(1),\ldots,p(n)} i^{|p|} u(1)^{p(1)} \ldots u(n)^{p(n)} M(p(1),\ldots,p(n))/p(1)!\ldots p(n)!$$

where

$$|p| \triangleq p(1) + \ldots + p(n) \text{ and}$$

$$M(p(1),\ldots,p(n)) \triangleq E(Z(t(1))^{p(1)} \ldots Z(t(n))^{p(n)}) \text{ is} \tag{19}$$

the centered moment of order p(1), ..., p(n) $\in \mathbb{N}$.

Similarly,

$$\psi(u) = \log \phi(u) \tag{20}$$

$$\triangleq \sum_{p(1),\ldots,p(n)} i^{|p|} u(1)^{p(1)} \ldots u(n)^{p(n)} \chi(p(1),\ldots,p(n))/p(1)!\ldots p(n)!$$

At this point, we define the cumulants $\chi$. By partial differentiation of $\psi(u) = \log \phi(u)$ with respect to u(1), it is easy to obtain the cumulants $\chi$ from the moments M by recursive linear equations.

The case n = 2 is of special interest: for intergers r and s,

$$\chi(r,s) = M(r,s) - \sum_{p,q \in D(r,s)} C_r^p\ C_{s-1}^q\ M(p,q)\ \chi(r-p,\ s-q) \tag{21}$$

with

$$D(r,s) \triangleq \{0,1,\ldots,r\} \times \{0,1,\ldots,s-1\} - \{(0,0)\}.$$

Putting $s = 0$, we deduce the case $n = 1$

$$\chi(r) = M(r) - \sum_{p=1}^{r} C_r^p \, M(p) \, \chi(r-p). \tag{21}$$

In this model, the cumulants $\chi$ are the essential tools because they can be computed readily from the observations and they are moments of a singular measure on $(\mathbb{R}^{n+1}, R^{n+1})$. This last fact comes from (17). As

$$\psi(u) = \int (\exp ia <u,G> - 1 - ia <u,G>)/a^2 \, d\lambda(a,s)$$

$$= \sum_{p(1),\ldots,p(n)} i^{|p|} (u(1)^{p(1)} \ldots u(n)^{p(n)}/p(1)! \ldots p(n)!)$$

$$\cdot \int G(t(1)-s)^{p(1)} \ldots G(t(n)-s)^{p(n)} \, a^{|p|-2} \, d\lambda(a,s).$$

By comparing coefficients of u,

$$\chi(p,(1),\ldots,p(n)) = \int G(t(1)-s)^{p(1)} \ldots G(t(n)-1)^{p(n)} a^{|p|-2} d\lambda(a,s) \tag{22}$$

Defining the mapping T by

$$T \overset{\Delta}{:} (a,s) \to (a, x(1) \triangleq G(t(1)-s), \ldots, x(n) \triangleq G(t(n)-s)$$

and letting $\lambda_T$ be the image measure of $\lambda$ under T, we get

$$\chi(p(1),\ldots,p(n)) = \int_{\mathbb{R}^{n+1}} x(1)^{p(1)} \ldots x(n)^{p(n)} \, d\lambda_T(x(1),\ldots,x(n)).$$

Whether or not the measure is singular, the measure $\lambda_T$ is positive so the cumulants $\chi$ have to satisfy the moment inequalities. If $\lambda_T$ is not positive, the model does not fit the experimental data.

Now we give details of the relation between the model and the inequalities. The moment inequalities on cumulants characterize a more general model than the previous one as will be explained.

Instead of one filtered Levy process, suppose that Z is the sum of K independent filtered Levy processes. With the obvious notation,

$$Z(t) = \sum_{\ell=1}^{K} \int_{-\infty}^{t} G(\ell, t-s) \, dL(\ell,s) \tag{23}$$

with the second characteristic function:

$$\psi(\langle f,\ Z\rangle) = \tag{24}$$

$$\sum_{\ell=1}^{K} \int_{\mathbb{R}^2} (\exp ia \langle f,G(\ell)\rangle - 1 - ia \langle f,G(\ell)\rangle)/a^2 \, d\lambda(\ell,\ a,\ s).$$

Defining the mapping $T(\ell)$ by

$$T(\ell) \overset{\Delta}{:} (a,s) \to b \overset{\Delta}{=} a \langle f,G(\ell)\rangle \overset{\Delta}{=} a \int_{s}^{\infty} f(t)\ G(\ell,\ t-s)\ dt,$$

a measure $\gamma$ exists on $(\mathbb{R}, \mathcal{R})$ such that for all real scalars v,

$$\psi(v) \overset{\Delta}{=} \psi(\langle vf,\ Z\rangle) = \int (\exp ivb - 1 - ivb)/b^2 \ d\gamma(b). \tag{25}$$

More precisely, we need to come back to definition (12). We omit the index $\ell$. Then

$$\psi(\nu) = - \int v^2\langle f,G\rangle^2 \ d\nu(s)/2 + \int (\exp iav \langle f,G\rangle - 1 - iav\langle f,G\rangle)\ d\mu(a,s)$$

$$= - \int v^2\langle f,G\rangle^2 \ d\nu(s)/2 + \int (\exp ivb - 1 - ivb)\ d\mu_T(b)$$

$$= \int (\exp ivb - 1 - ivb)/b^2 \ \{\int\langle f,G\rangle^2 \ d\nu(s)\ \delta(db) + b^2 \ d\mu_T(b)\}.$$

The singularity $\sum_{\ell} \int \langle f,G(\ell)\rangle^2 \ d\nu(\ell,s)\ \delta(db)$ of the measure depends on the function f.

In the sea, the Levy process is spread over the spatial temporal space and the causal linear filter G depends on the relative position of the sources and sensors. We should write

$$p(x,\ y,\ z,\ t) = \int G(x-\xi,\ y-\eta,\ z-\zeta,\ t-\tau)\ dL(\xi,\ \eta,\ \zeta,\ \tau).$$

Moreover, the impulse response G may be random although sources and response G are stochastically independent.

We can show by a similar argument that for the general case of random linear filtering of temporal Levy processes, the observation Z has the second characteristic function:

$$\psi(v) = \log E\ (\exp iv \langle f,Z\rangle) \tag{26}$$

$$= \int_{\mathbb{R}} \{([\exp ivb] - 1 - ivb)/b^2\}\ d\gamma(b).$$

If we introduce the cumulants $\chi(p)$, $p \in \mathbb{N}$,

$$\psi(v) = \sum_{p=2}^{\infty} (iv)^p \ \chi(p)/p! \quad \text{and} \tag{27}$$

$$\chi(p) = \int_{\mathbb{R}} b^{p-2} \ d\gamma/b). \tag{28}$$

The Hankel matrix

$$\chi \triangleq \begin{vmatrix} \chi(2) & \chi(3) & \cdots \\ \chi(3) & \cdots & \\ \cdots & & \end{vmatrix} \qquad (29)$$

is positive. Any positive measure's moments have this property.

The restrictive equalities (5) and (6) of the Gaussian process Z are replaced by inequalities: for $n \geq 1$ and $c(1), \ldots, c(n) \in \mathbb{C}$

$$\sum_{p,q} c(p)\, \overline{c(q)}\, \chi(p+q) \geq 0.$$

With our actual sea observations, taking account of the statistical error estimation, sometimes the restrictive equalities (5) and (6) are fairly well satisfied for durations of less than one second. For larger durations energy changes and spherically invariant models (4-3) would be necessary. For short durations, the Levy model improves the goodness of fit and a flickering effect is thus emphasized.

In Section 5, we examine identification of the measure $\gamma$ from its moments, using adaptations of the classical moments technique.

## 4. STATIONARY LEVY PROCESSES WITH SPHERICALLY INVARIANT JUMPS

In this section, we consider the Levy processes

$$dL(s) = dW(s) + \sum_{i \in Z} A(i)\, \delta(s(i), ds) \qquad (1)$$

in which the law of A is spherically invariant. We begin by recalling the definition of a spherically random variable Y and give some direct applications to models in underwater acoustics.

The spherically invariant characteristic function is given with a positive measure $\chi$ on $(\mathbb{R}_+, R_+)$ by the expression

$$\phi(u) \triangleq E(\exp(iuY)) = \int_{\mathbb{R}_+} \exp(-\sigma^2 u^2/2)\, d\chi(\sigma). \qquad (2)$$

As $\phi(0) = 1$ the measure $\chi$ is a probability on $(\mathbb{R}_+, R_+)$. For the approximation

$$\chi \triangleq \chi(1)\, \delta(\sigma(1), \cdot) + \ldots + \chi(n)\, \delta(\sigma(n), \cdot),$$

the characteristic function is equal to

$$\phi(u) = \sum_{\ell=1}^{n} \chi(\ell) \exp - (\sigma(\ell)\ u^2/2).$$

Thus Y is a convex mixture of n centered Gaussian variables with variances $\sigma(1)$, ..., $\sigma(n)$ respectively. The law of Y is also the law of the product AX where X is a centered normalized Gaussian variable, A is a random variable equal to $\sigma(\ell)$ with probability $\chi(\ell)$, and A, X are a pair of independent variables.

Generalization of these two representations is obvious for any probability $\chi$.

A good approximation of the law of the observation Z studied by C.R. Baker (see Chapter 6, this volume) is to suppose that

$$\begin{cases} Z(1) = A\ X(1) \\ Z(2) = A\ X(2) \\ \\ Z(q) = A\ X(q) \end{cases} \tag{3}$$

where $X \triangleq (X(1), \ldots, X(q))^*$ is a centered Gaussian vector with covariance $\Gamma \triangleq E\ (X\ X^*)$, and where $A^2$ represents a random energy factor.

If we put $Y \triangleq \langle u, Z\rangle = u(1)\ Z(1) + \ldots + u(n)\ Z(n)$, we obtain the characteristic function of Z:

$$\phi(u) = \int \exp - (\sigma^2\ u\Gamma u^*/2)\ d\chi(\sigma). \tag{4}$$

Here u is any vector in $\mathbb{R}^n$.

For a covariance matrix $\Gamma$ estimated or deduced by geometrical hypotheses, the measure $\chi$ may be approximated by moment techniques (§ 5). From equation (3-23) and the Fourier transform, we get the model

$$Z(\nu) = \sum_{\ell=1}^{K} G(\ell, \nu)\ L(\ell, \nu).$$

Because G describes the effect of the propagation, a product decomposition is assumed:

$$G(\ell, \nu) = A(\ell, \nu)\ g(\ell, \nu)$$

where the scalar A is random and g is a geometric term.

Finally

$$Z(\nu) = \sum_{\ell=1}^{K} A(\ell, \nu)\ X(\ell, \nu) \text{ where}$$

$$X(\ell, \nu) = g(\ell, \nu)\ L(\ell, \nu).$$

With the hypothesis that $X(\ell, \nu), \ldots, X(K, \nu)$ are independent and centered Gaussian r.v.'s conditioned by A, it is possible to identify the number K of sources [7, Chapter G-II).

We now return to Levy processes as in (1). Suppose the law of A is spherically invariant with probability $\chi$ and suppose the discrete point process $(s(i); i \in I)$ is stationary Poisson with density $\alpha$. Direct computation then gives

$$\psi(u) = \log E(\exp iu \langle f, Z\rangle) \tag{5}$$

$$= \int_{\mathbb{R}\times\mathbb{R}_+} ((\exp - (\sigma^2 u^2 \langle f, G\rangle/2) - 1)/\sigma^2)\, \alpha(\sigma_o^2\, \delta(d\sigma) + \sigma^2\, d\chi(\sigma))\, ds.$$

We observe the stationary Gaussian part given by $\sigma_0^2\, \alpha\delta(d\sigma)\, ds$.

We have to show that expression (5) of the characteristic function is indeed a particular case of (3-13).

<u>Proposition</u>: If

$$d\lambda(a,s) \triangleq \alpha(\sigma_0^2\, \delta(da) + \int \exp - (a^2/2\sigma^2)/\sqrt{2\pi\sigma}\, d\chi(\sigma)\, a^2 da)ds, \tag{6}$$

then $\psi(u)$ defined in (5) is given by

$$\psi(u) = \int (\exp iua \langle f, G\rangle - 1 - iua \langle f, G\rangle)/a^2\, d\lambda(a,s).$$

Proof:

$$\int [(\exp iua \langle f, G\rangle - 1 - iua\langle f, G\rangle)/a^2\, (\int \exp - (a^2/2\sigma^2)/\sqrt{2\pi\sigma}\, d\chi(\sigma))]\, a^2 da$$

$$= \int d\chi(\sigma) \int ((\exp iua\langle f, G\rangle - 1 - iua\langle f, G\rangle)/\sqrt{2\pi}\, \sigma) \exp - (a^2/2\sigma^2)\, da$$

$$= \int d\chi(\sigma)\, (\exp - (u^2\sigma^2 \langle f, G\rangle^2/2) - 1).$$

□

As filtered Levy Processes with spherically invariant jumps are still filtered Levy Processes, the technique of identification by cumulants of measure $\lambda$ is practical but we prefer direct identification of $\alpha(\sigma_o^2\, \delta(d\sigma) + \sigma^2\, d\chi(\sigma))\, ds$ as in (5).

The equalities of coefficients for $u^\ell$, $\ell \in \mathbb{N}$ are written:

$$\begin{cases} \chi(2) = 1 = \alpha\int \langle f, G\rangle^2\, ds\, (\sigma_o^2 + \chi(\mathbb{R}_+)) \\ \chi(2p) = ((2p!)/2^p\, p!) \int \langle f, G\rangle^{2p}\, \alpha\, ds \int \sigma^{2p-2}\, d\chi(\sigma) \\ \chi(2p+1) = 0. \end{cases} \tag{7}$$

By hypothesis we normalize the observation: $\int \langle f, G\rangle^2\, \alpha\, ds = 1$. Thus $\sigma_o^2\, \delta(d\sigma) + \sigma^2\, d\chi(\sigma)$ is a probability.

## 5. MEASURE IDENTIFICATION BY THE METHOD OF MOMENTS

The classical theory proceeds as follows.

Given 2n moments m(0), ..., m(2n-1) of a positive bounded measure $\nu$, we consider the approximation

$$d\hat{\nu}(b) \triangleq \nu(1)\ \delta(b(1),\ db) + \ldots + \nu(n)\ \delta(b(n),\ db) \tag{1}$$

which has to satisfy

$$\begin{aligned} m(\ell) &= \nu(1)\ b(1)^{\ell} + \ldots + \nu(n)\ b(n)^{\ell} \\ &= \int b^{\ell}\ d\nu(b) = \int b^{\ell}\ d\hat{\nu}(b) \qquad \ell = 1,\ldots,n. \end{aligned} \tag{2}$$

These 2n equations are linear in $\nu(1)$, ..., $\nu(n)$ and algebraic in b(1), ..., b(n). The solution is known [1, Chapter ]].

We consider the Hilbert space $L^2$ ($\mathbb{R}$, $\mathcal{R}$, $d\nu$) and make the Gram-Schmidt orthonormalization for the n + 1 polynomials:

$b \to b^{\ell}$  $\ell = 0, 1, \ldots, n$. Thus we obtain n + 1 orthonormal polynomials

$$P(\ell) \overset{\Delta}{:} b \to P(\ell,b) = p(\ell,0) + p(\ell,1)\ b + \ldots + p(\ell,\ell)\ b^{\ell}.$$

The n zeros b(1), ..., b(n) of $P(\ell)$ are precisely the b solution of equations (2); then $\nu(1)$, ..., $\nu(n)$ are the solution to the first n linear equations.

The approximation (1) seems too rough in many cases, especially if the measure $\nu$ is known to be absolutely continuous with respect to Lebesgue measure. A better approximation would be

$$\begin{aligned} d\nu(b) &= (\mu(1)\ \delta(b(1),\ db) + \ldots + \mu(n)\ \delta(b(n),\ db) \\ &\quad * H\ db \\ &= (\mu(1)\ H(b-b(1)) + \ldots + \mu(n)\ H(b-b(n)))\ db \end{aligned} \tag{3}$$

where H(b) db is a given probability with density H. The moment equations to be solved in $\mu(1)$, ..., $\mu(n)$ and b(1), ..., b(n) are

$$m(\ell) = \sum_{k=1}^{2n-1} \mu(k) \int b^{\ell}\ G(b-b(k))\ db. \tag{4}$$

To obtain the solution, consider the convolution product

$$\nu = \mu * H\ db \text{ of two measure } \mu \text{ and } H\ db. \tag{5}$$

By the Fourier transform,

$$\phi(\nu,\ u) \triangleq \int \exp\ (iub)\ \nu(db) = \phi(\mu,\ u)\ \phi(H\ db,\ u) \qquad (6)$$

for every u in $\mathbb{R}$.

We need only 2n moments of the measure $\mu$ and they are solutions to 2n linear and recursive equations. We develop the two members of (6), denoting $m(\nu,\ p) \triangleq \int b^p\ \nu(db)$. Then

$$\sum_{\ell=0}^{\infty} (iu)^{\ell}\ m(\nu,\ell)/\ell! = \sum_{p=0}^{\infty} (iu)^p\ m(\mu,\ell)/p!\ \sum_{q=0}^{\infty} (iu)^q\ m(H\ db,\ q);$$

hence

$$m(\nu,\ell) = \sum_{p+q=\ell} ((p+q)!/p!\ q!)\ m(\mu,p)\ m(H\ db,\ q).$$

Thus,

$$m(\mu,\ell) = m(\nu,\ell) - \sum_{p=0}^{\ell-1} C_{\ell}^{p}\ m(\mu,p)\ m(H\ db,\ \ell-p).$$

The Laplace transform of a measure $\nu$ is a better way of obtaining an accurate approximation. By hypothesis, for a random variable B with probability measure $\nu$

$$\int e^{-\lambda b}\ \nu(db) \triangleq E\ (\exp - \lambda B) \cong \frac{1}{N} \sum_{k=1}^{N} \exp - \lambda B(k) \qquad (7)$$

where $(B(k),\ k = 1,\ \ldots,\ N)$ are i.i.d. as B.

A change of variable $T\colon\ b \to c = \exp - b\Delta\lambda$ gives the image probability $\nu_T(dc)$ and the equality

$$\int (\exp - b\Delta\lambda)^{\ell}\ \nu(db) = \int c^{\ell}\ \nu_T(dc).$$

Here again we have a moment problem for the measure $\nu_T$. If

$$\nu_T(db) \triangleq \sum_{k=1}^{n} \nu(k)\ \delta(c(k),\ dc),\ \text{then}\ \nu(db) = \sum_{k=1}^{n} \nu(k)\ \delta(\log c(k)/-\Delta\lambda,\ db).$$

When $\Delta\lambda$ is small the number 2n of moments available is large; it is sufficient that $2n\ \Delta\lambda < 1$ for a good estimate.

For a normalized Gaussian random variable B, we have $E(\exp - B(k)\ \Delta\lambda\ell) = \exp \frac{1}{2}(\Delta\lambda\ell)^2$ and the variance of the estimate (7) is

$$\sigma_{\ell}^2 = (\exp 2(\ell\Delta\lambda)^2 - \exp (\ell\Delta\lambda)^2)/N.$$

For $\Delta\lambda = 0.01$, the number $2n = 100$ is easily used to approximate $\nu_T$ with fifty terms.

Application to the Levy process is easy because of the relation between Laplace and Fourier transforms; setting $\lambda = -iu$ gives $E(\exp - \lambda B) = E(\exp iuB)$.

Starting, for example, from (3-25)

$$\log E(\exp - \lambda\langle f, Z\rangle) = \int_{\mathbb{R}} ((\exp - \lambda b - 1 + \lambda b)/b^2)\, d\gamma(b).$$

A polynomial approximation for small $\lambda$,

$$\log E(\exp - \lambda\langle f, Z\rangle) = \sum_{\ell=2}^{2n+1} (-\lambda)^{\ell}\, \chi(\ell)/\ell!$$

with equality if $\lambda = \ell\Delta\lambda$, $\ell = 0, \ldots, 2n-1$, gives the required cumulants directly.

## 6. STATIONARY CASE: IDENTIFICATION OF THE IMPULSE RESPONSE

Extra hypotheses are necessary to estimate the moment of the centered normalized second order process Z from a trajectory. We suppose the process Z to be ergodic which implies for the measure $\lambda$ of the Levy process the product

$$d\lambda(a,s) = \sigma(da) \otimes ds.$$

The cumulants are (3-22)

$$\begin{aligned}\chi(p,q) &= \int G(t(1)-s)^p\, G(t(2)-s)^q\, a^{p+q-2}\, d\lambda(a,s) \\ &= \int G(t(1)-s)^p\, G(t(2)-s)^q\, ds \int a^{p+q-2}\, \sigma(da)\end{aligned} \tag{1}$$

and are products of the two factors $I(p,q) \triangleq \int G(t(1)-s)^p\, G(t(2)-s)^q\, ds$ and $\int a^{p+q-2}\, \sigma(da)$.

To eliminate the measure $\sigma$, we consider the quotient $\Gamma(p,\tau) = A_1/A_2$,

$$A_1 \triangleq \int G(t(1)-s)^p\, G(t(2)-s)^p\, ds$$

$$A_2 \triangleq \left|\int G(t(1)-s)^{2p}\, ds \cdot \int G(t(2)-s)^{2p}\, ds\right|^{\frac{1}{2}}.$$ This can be written as

$$\begin{aligned}\Gamma(p,\tau) &= \int G(r)^p\, G(r-\tau)^p\, dr \Big/ \int G(r)^{2p}\, dr \\ &= \chi(p,p) / \left|\chi(2p,0)\, \chi(0,2p)\right|^{\frac{1}{2}}\end{aligned} \tag{2}$$

where $\tau \triangleq |t(2) - t(1)|$.

We see that $\Gamma(p,\cdot)$ has the properties of a covariance: $\Gamma(p,0) = 1$ and $\Gamma(p,\cdot)$ is positive definite.

When $G$ is continuous, we obtain by Lebesgue's theorem the continuity of $\Gamma$. Thus $\Gamma(p,\cdot)$ defines a spectral measure $F(p,\nu)$.

If $\int |G(r)|^p \, dr < \infty$ then we introduce the Fourier transform

$$G(p, \nu) \triangleq \int \exp(-i\nu r)\, G(r)^p \, dr/2\pi \tag{3}$$

and

$$G(r)^p = \int \exp(i\nu r)\, G(p, \nu)\, d\nu.$$

As $\int |G(r)|^{2p} \, dr$ is finite, using the Fubini theorem in (1) gives

$$\Gamma(p,\tau) \triangleq \int \exp(i\nu\tau)\, dF(p,\nu) \tag{4}$$

$$= \int \exp(i\nu\tau)\, |G(p,\nu))^2 \, d\nu / \int |G(p,\nu)|^2 \, d\nu.$$

The spectral measure $F(p,\nu)$ is absolutely continuous with respect to $d\nu$ and

$$dF(p,\nu) = |G(p,\nu)|^2 \, d\nu / \int |G(p,\nu)|^2 \, d\nu.$$

The case $p = 1$ is of special interest. We omit the index $p$ in the following formulas. To begin, the normalizations $E(Z(t)) = 0$ and $E(Z(t)^2) = 1$ give

$$\chi(2, 0) = \chi(0, 2) = \int G(r)^2 \, dr \int d\sigma(a) = 1.$$

We suppose $\int G(r)^2 \, dr = 1$ thus $\sigma(\mathbb{R}) = 1$.

The function $\Gamma$ is in fact a covariance:

$$\Gamma(\tau) \triangleq \chi(1, 1) = E(Z(t)\, Z(t-\tau)) = \int \exp(i\nu\tau)\, |G(\nu)|^2 \, d\nu. \tag{4}$$

We have to determine $G$ from $\Gamma$ using (4): as this problem is well known in control theory, we simply recall the results. We have to determine the impulse response $G(t)$, $t \in \mathbb{R}$ on the frequency response $G(\nu)$, $\nu \in \mathbb{R}$ from the covariance of an output of a $G$ linear filter excited by a white noise.

In practice, the time is discretized with unit $\Delta t$. Moreover the filter is supposed to be rational:

If $(e(t);\ t \in \mathbb{Z})$ is a scalar white noise then

$$Z(t) + a(1)\, Z(t-1) + \dots + a(p)\, Z(t-p) = b(0)\, e(t) + \dots + b(q)\, e(t-q)$$

is the ARMA model of the filter with transfer function $\forall\, z \in \mathbb{C}$

$$H(z) \triangleq N(z)/D(z)$$

$$N(z) \triangleq b(0) + b(1)\, z^{-1} + \dots + b(q)\, z^{-q}.$$

$$D(z) \triangleq 1 + a(1)\, z^{-1} + \dots + a(p) z^{-p}. \tag{5}$$

If $q = 0$ one speaks of an auto regressive model.

If $p = 0$ one speaks of a moving average model.

The causal filter assumes the zeros of $D(z)$ to be inside the unit circle and the phase minimal filter assumes the zeros of $N(z)$ to be inside the unit circle. As

$$|G(\nu)|^2 = H(\exp i\nu)\, H(\exp - i\nu)$$

is estimated (4), it is possible, for every p and q, to compute $b(0), \ldots, b(q)$ and $1, a(1), \ldots, a(p)$ so that the filter is causal and phase minimal.

With the experimental results, we use the Itakura algorithm to get the causal auto-regressive coefficients $1, a(1), \ldots, a(p)$ [11].

## 7. CONCLUSION

In going from underwater pressure to the analysis of phenomena, numerous steps are necessary. Our pressure data were measured near the South coast of France by a naval laboratory. A magnetic tape with 14 analog tracks out of 16 was sent to Rouen.

The 14 tracks were then sampled simultaneously several times. Three main programs in Pascal for the discretized signals Z were written, as now described.

First, each of the $q = 14$ coordinates of the observation Z is centered and normalized. Next, moments up to order twenty and co-variances with twenty lags are recursively computed. Recursion is used to decrease computational errors and avoid memory problems [2].

Next, the measure identification of § 5 is made. From the moments, we compute successive orthonormal polynomials and for the last one, its zeros $b(1), \ldots, b(n)$.

Finally, the weights $\nu(1), \ldots, \nu(n)$ are obtained from solving a linear system.

To move from the observed moments $M(\ell)$ of the observation Z (2-19) to the moments $m(\ell)$ of a measure $\nu$ (5-2) many operations will occur, depending on the model used: spherically invariant (4-3), Levy process (3-28), Levy process with spherically invariant jumps (4-7), stationary class (6-2).

In the stationary cases, we use another algorithm to estimate an impulse response G from a covariance. The A.R. coefficients are computed by the Itakura algorithm and for each order p, modeling error is estimated until this error no longer decreases.

We check that the zero of the associated polynomial (6-5) is inside the unit circle. The impulse response G(t) is the Fast Fourier Transform of $G(\nu)$ and the time integrals in (6-1) are computed by summation.

For treatment with a microcomputer, we split up the program into twelve programs of about two hundred lines.

The spherically invariant model and the Levy process model improved the fit to our experimental data. This indicates the interest of such models.

The main result for the study of 2000 values taken every $\Delta t = 1.25\ 10^{-4}$ sec from hours of recordings is the good fit between model and experimental data: we consider the stationary case of section 6. We estimate moments and covariance. We deduce the impulse response G point by point; roughly it looks like sin t/t; only two delays in the A.R. model are necessary.

The moments do not satisfy Gaussian moments equalities and the statistical law is not symmetric. We consider (6-1):

$$\chi(p) = \int G(t-s)^p\, a^{p-2}\, d\lambda(a,s) = \int G(t-s)^p\, ds \int a^{p-2}\, \sigma(da)$$

and compute the Hankel matrices for $\sigma$(3-29). They decrease quickly to zero: 1, $10^{-2}$, $10^{-5}$ and so on. Development of the second characteristic function in $u^2$, $u^3$ and $u^4$ is sufficient.

This improvement indicates that additional work should be done, not only to verify these results with additional data, but also to study the dependence of the model parameters on various physical parameters such as sea state, location, etc.

NOTATIONS INDEX

$\triangleq$ equal by definition

* adjoint (example: A* adjoint matrix or transpose complex conjugate matrix) or convolution (as in f*g)

log logarithm to base e

$\Delta_p = (\frac{\partial^2}{\partial x^2} + \frac{\partial^2}{\partial y^2} + \frac{\partial^2}{\partial z^2})\, p$ Laplacian of p

$\otimes$ product of measurable spaces

$L^2$ $(\mathbb{R}^4, R^4, dx\, dy\, dz\, dt)$ Hilbert space of real function $\phi$ of the four variables x, y, z, t measurable and square integrable with respect to the Lebesgue measure dx dy dz dt in $(\mathbb{R}, \mathcal{R})^{\otimes 4}$

$q! = q \cdot (q-1) \cdot \ldots \cdot 3 \cdot 2 \cdot 1$   factorial q

$|p| = p_1 + p_2 + \ldots + p_n$

$\langle f, Z \rangle \triangleq \int_{\mathbb{R}} f(t)\, Z(t)\, dt$

$\langle f, G \rangle(s) \triangleq \int_s^{\infty} f(t)\, G(t-s)\, dt$

$\langle u, Z \rangle \triangleq \sum_{i=1}^{n} u(i)\, Z(t(i))$

c   phase velocity

k   wave number

q   number of sensors: the observation is q dimensional

$Y(t) = 1_{[0,\infty[}(t)$   Heaviside function

$\alpha$   attenuation coefficient

$\delta(s(i), ds) = \delta_{s(i)}(ds)$   Dirac measure at point s(i)

ACKNOWLEDGEMENTS

The § 2 research was supported by French Cethedec Contracts and § 4 by ONR Contract N00014-81-K-0373.

I am greatly indebted to Professor Charles Baker for numerous suggestions and for corrections on the manuscript.

REFERENCES

1. N.I. Akhiezer, The Classical Moment Problem, University Mathematical monographs, Oliver & Boyd LTD (1965).

2. T.F. Chan, G.H. Golub and R.J. Leveque, Updating formulae and a pairwise algorithm for computing sample variance, Compstat Proceeding in Computational Statistics, 30-41, Physica Verlag (1982).

3. S. Chernov Lev, Wave Propagation in Random Medium, Dover Publications, Inc., New York (1967).

4. D. de Brucq, Filtrage linéaire et propagation acoustique, 6éme Congrès Gretsi, 3/1-3/6 (1977).

5. D. de Brucq, Modelisation d'un enregistrement de bruit marin et détection, Colloque Gretsi, 12/1-12/7 (1979).

6. D. de Brucq et C. Olivier, Approximations des processus Gaussiens stationnaires, solutions d'équations aux dérivées partielles linéaires, Rev. Roum. Math. Pures et Appl., Tome XXVIII n° 3, 205-228, Bucarest (1983).

7. D. de Brucq, Cours de Théorie du Signal, D.E.A. de Mathématiques Appliquées de Rouen, Rouen (1984).

8. Gikhman et Skorokhod, Theory of Random Processes, Saunders Co. (1969).

9. Guyesse et Sabathe, Acoustique sous-marine, Dunod, Paris.

10. T. Hida, Stationary Stochastic Processes, Mathematical Notes, Princeton University Press, New Jersey (1970).

11. F. Itakura, S. Saito, Digital filtering techniques for speech analysis and synthesis in Proc., 7th Int. Cong. Acoust., 261-264, Budapest (1971).

12. C.C. Leroy, Problèmes liés au milieu marin lors de l'utilisation des fréquences ultrasonores, 3e Congrès Gretsi (1971).

13. H. Mineur, Techniques de Calcul Numérique, Dunod, Paris.

14. B. Picinbono, Spherically invariant and compound Gaussian stochastic processes, IEEE Trans. on Information Theory, I.T.-6, 77-79 (1970).

15. B. Poiree, Les équations de l'acoustique non linéaire dans les fluides dissipatifs, Revue du Cethedec, vol. 1, n° 46 (1976).

16. L. Schwartz, Méthodes mathématiques de la physique, Convolution III, Centre de Documentation Universitaire, Edition S.E.D.E.S. (1960).

17. G. Ruckebush, Sur le problème de la synthèse des filtres, Colloque CNRS, Aussois, 21-25 Sept., vol. 2, A1-1, A1-21 (1981), INRIA BP 105 78153 Le Chesnay, France.

18. M. Westcott, Identifiability in linear processes, Z. Wahrscheinlichkeiststheorie verw Gerb., 16, 39-46 (1970).

19. Fast algorithms for linear dynamical systems, Aussois, 21-25 Sept. 1981 - INRIA.

CHAPTER 2

# MULTIPLE TIME DELAY ESTIMATION IN UNDERWATER ACOUSTIC PROPAGATION

**G. Jourdain and M.A. Pallas**

## 1. INTRODUCTION

This chapter is essentially devoted to the active IDENTIFICATION of a propagation channel between an emitter and a receiver. Our example deals with the underwater channel, though this remains valid for any kind of propagation channel (whatever the nature of the transmitted wave -electric, acoustic, seismic- may be) and for applications to various domains (vibrations, industrial processes, biology, thermoanalysis) in addition to the classical ones (radar, sonar, automatics, telecommunications...).

The only basic hypothesis made on the channel is LINEARITY with respect to transmitted signals. This assumption is very little restrictive. No additional information or model is assumed ; so it is an a priori NON PARAMETRIC identification.

System terminology is used, that is to say the transmission channel is represented by a linear system establishing a relationship between one input and one output (bipunctual system), or N inputs and M outputs (such as antennas in underwater acoustics). The system is then characterized by its RESPONSE to the emitted signal. According to more or less rapid system time variations, this response may be stationary or not. This description also enables us to take account of statistical or random channel aspects (this is sometimes necessary for example in the case of ionospheric or tropospheric channels, underwater acoustics, urban communication channel [3,4,7,11,17,20]). Of course, the channel response is perturbed by noise between emitter and receiver.

In underwater acoustics, solving wave propagation equations is an usual way of representing acoustic propagation. According to different hypothesis on geometric conditions, distance, bottom and surface, and various simplifications, one adopts either more or less elaborated RAY THEORY, or mode theory, or parabolic equation... These descriptions are quite essential, but on the one hand they are approximated, and often give a rather energetic aspect of the transmitted wave. On the other hand, they correspond primarily to a STATIC description of the channel. The proposed identification is complementary. It allows us to better take account of variational, evolutive and/or statistic propagation aspects.

It is clear that in undersea acoustics, the first phenomenon is MULTIPATH propagation of the acoustic wave. It is particularly exhibited in ray tracing. In our identification process, these different paths will appear and the method which starts as a non parametric identification can become PARAMETRIC by estimating the characteristic parameters of the different paths.

Let us see now what the possible objectives in identifying a system are, because the objective must or can guide the identification and define its limits. A first purpose may be the knowledge of the system itself. In underwater acoustics, this geophysic aim is proceeded by oceanologs and physicists. Let us recall for example the recent and future developments of acoustic tomography [6,18]. A second and more operational objective consists in getting some knowledge on the channel : underwater applications are for example prospection (petroleum, ore), and sonar of course (detection, communication, antenna imaging...). In our case, undersea channel identification is rather applied to underwater acoustic communications [12].

This chapter includes 3 parts : the first one develops the methodology used for active nonparametric identification of a channel, the suitable emitted signals and the corresponding processing. The second part is concerned with the structure and performance of multipath estimation in underwater acoustics ; some elements of basic parameter estimation theory are recalled. And the third part gives a precise underwater example in a case of multipath propagation and presents some results.

## 2. METHODOLOGY OF NONPARAMETRIC IDENTIFICATION

### 2-1. THE TRANSMISSION MODEL

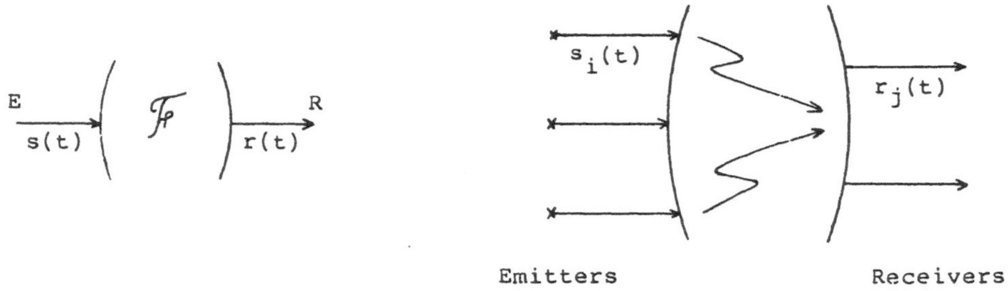


FIGURE 1

Let us consider the linear system F whose input is s(t), and output r(t). In the multiple input-output case, there are as many defined systems as input-output couples. The system F is assumed LINEAR with respect to the emitted signals s(t). Then in the most general way, F is characterized by its BITEMPORAL IMPULSE RESPONSE (i.r.) $H(t,\varphi)$ which may be time varying :

$$r(t) = \int H(t,\varphi)\; s(t-\varphi)\; d\varphi \tag{1}$$

It is well known that this linear filter can be equivalently characterized by any of the other three responses obtained from $H(t,\varphi)$ by taking a Fourier transform with respect to one or other variable t or $\varphi$ [1,13].

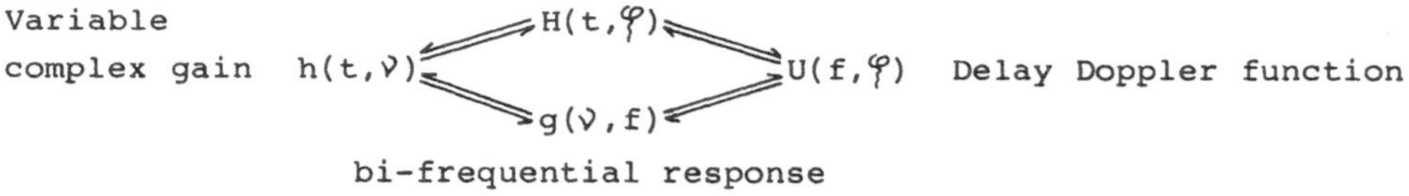


Of course, noise must be taken into account over the transmission E-R. The simplest following model is then adopted, with additive random noise b(t)

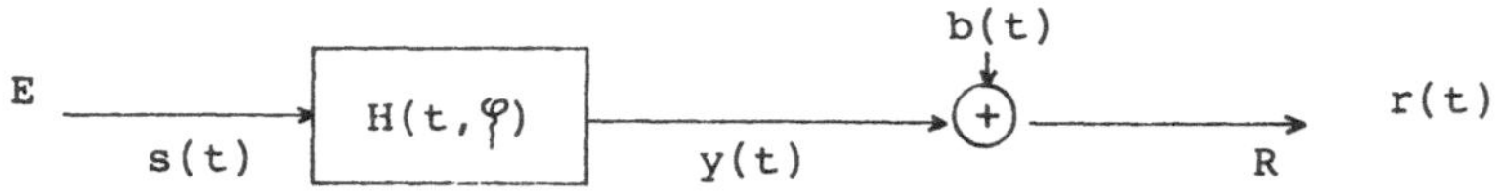


FIGURE 2 : Transmission model

## 2-2. NON PARAMETRIC IDENTIFICATION

2-2-1. With this general model and by assuming in a first step that $H(t,\varphi)$ is time invariant (i.e. $H(\varphi)$ only), the optimal identification method of H, when using a second order criterion, is the following [9] : one looks for the i.r. $H(\varphi)$ which minimizes the mean square error between the received signal r(t) and the convolution $H * s(t)$

$$\min_H \quad E\left\{ | r(t) - (H * s)(t) |^2 \right\} \tag{2}$$

NOTE : In the above expression, the filter H is assumed deterministic, and s(t) and r(t) are random, stationary, independent of b(t). If signals s(t) and y(t) are non random, a quadratic temporal mean criterion is used.

The solution of (2) is well known (orthogonality principle) : H must satisfy the following equation

$$\Gamma_{rs} = H * \Gamma_s \tag{3}$$

where $\Gamma_s$ and $\Gamma_{rs}$ are the auto and mutual stationary covariance functions of s(t) and r(t). (In the non random case, they are auto and intercorrelation functions).

2-2-2. Let us assume now that F is time varying and let us generalize this identification method. By looking at Figure 2, even if s(t) is stationary, y(t) and r(t) will not be stationary, because of the time variation of $H(t,\varphi)$ (except for special cases). One always looks for the i.r. $H(t,\varphi)$ which minimizes $E\left\{| r(t) - y(t) |\right\}^2$ , where y(t) is the generalized convolution (1). Orthogonality principle leads to orthogonality between s(t) and the difference r(t) - y(t), so that H must satisfy :

$$E\left\{ r(t)\ s^*(u) \right\} = E\left\{ \int H(t,\varphi)\ s(\varphi)\ s^*(u)\ d\varphi \right\}$$

so

$$\Gamma_{rs}(t,u) = \int H(t,\varphi).\ \Gamma_s(\varphi - u)\ d\varphi \tag{3'}$$

which is the obvious generalization of (3).

2-2-3. So in all cases, the intercorrelation $\Gamma_{rs}$ and the autocorrelation $\Gamma_s$ are involved in the identification of H. If the emitted signal s(t) is chosen in order that its autocorrelation $\Gamma_s$ looks like a Dirac function :

$$\Gamma_s(\tau) \sim \gamma_0 \ \delta(\tau) \tag{4}$$

then the intercorrelation $\Gamma_{rs}$ will provide directly the i.r. H, except for a constant.

2-2-4. All this can be extended to the case when F itself becomes a random filter. Some authors [1,13,17] have studied this case. It may sometimes be the right model in underwater acoustics : it depends on the rapidity of variation of F with respect to the time range of interest. Then the i.r. $H(t,\varphi)$ (and all other responses) must be considered as random processes. The characterization of F is made by means of STATISTICAL PROPERTIES of these processes. This leads to different classes of random filters, one of them being very often retained : the WSSUS (Wide Sense Stationary Uncorrelated Scatterer) filter, where the covariance of $H(t,\varphi)$ is stationary with respect to t, and uncorrelated with respect to $\varphi$ . The statistic description of F is then resumed in the channel "scattering function" [13]. This case will no longer be developed here ; in the whole following the i.r. H will be assumed non random.

## 2-3. MULTIPATH CHANNEL

The more precisely interesting case here, which is very often encountered in underwater acoustics is multipath channel. This model is also very often used in ionospheric propagation, and in urban communication too. Ray theory in sea propagation leads the acoustic energy to follow some privileged paths, and from an emitter point E to a receiver point R the acoustic wave is transmitted by multiple paths. An example is given in Figure 3 (from [16]), where one can see the sound celerity profile c(z) and the corresponding rays along which the acoustic wave is propagated for a given situation of E.

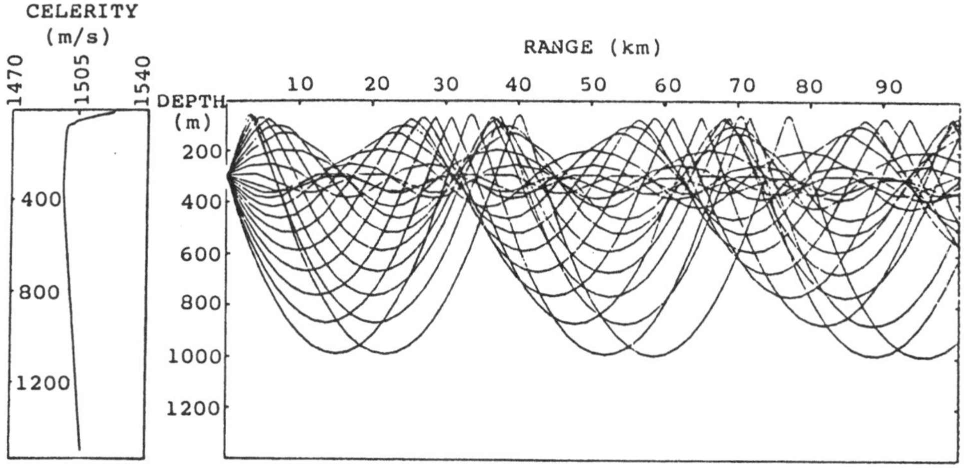


FIGURE 3 : Example of ray tracing. (from [16])

2-3-1. Ideal multipath case.

In this case, the i.r. becomes :

$$H(t, \varphi) = \sum_i \alpha_i(t) \,.\, \delta(\varphi - \tau_i(t)) \quad (5)$$

so that y(t) is

$$y(t) = \sum_i \alpha_i(t)\; s(t - \tau_i(t))$$

This is illustrated in Fig. 4 where the attenuations $\alpha_i$ and delays $\tau_i$ can be time varying (they could even be randomly varying in a more general model).

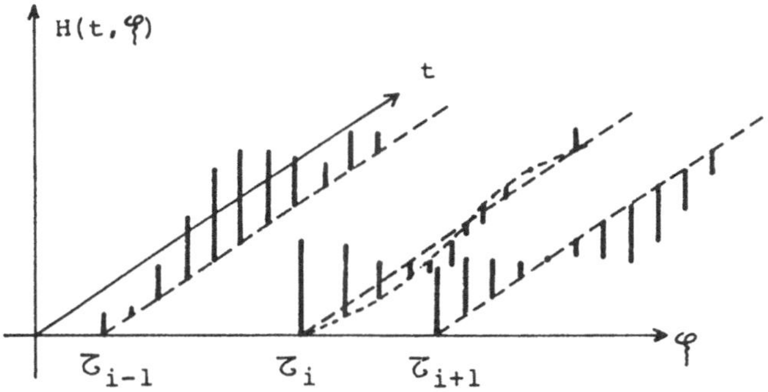


FIGURE 4 : Ideal Multipath Channel

An experimental result corresponding to $(H(t,\varphi))^2$ is given in Fig. 5 (from [8]).

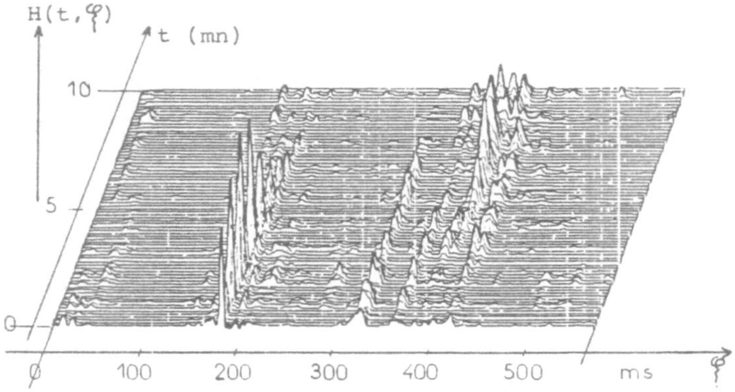


FIGURE 5 : Experimental Multipath channel

The multipath phenomenon is well exhibited and according to the time scale of interest, a temporal variation of transmitted energy over each path may occur (fading).

So, if the model (5) is adopted, the problem is no longer to identify H, but the parameters $\alpha_i$, $\tau_i$ (possibly time varying). And the non parametric identification of H is replaced by a parametric estimation.

In the following, given the time scale of interest for us (for example some seconds, or minutes for a communication message), the parameters $\alpha_i$ and $\tau_i$ are stable enough and are assumed CONSTANT.

2-3-2. Experimental multipath case.

An important point in underwater acoustics is that all signals -and the undersea channel too- are bandpass. The notation of modulated signals round a carrier frequency $\nu_0$ is then used, and the channel has also to be characterized over a bandpass frequency range round $\nu_0$. It turns out that the model (5) must be completed in order to take account of a phase displacement of the carrier frequency, which is always present and possibly different from one path to another. The corresponding model for the i.r. of F is often written by using the signals and i.r. complex amplitudes, relative to $\nu_0$. For example, a possible way consists in assuming that $\alpha_i(t)$ can be complex quantities :

$$\alpha_i(t) = \rho_i(t)\ e^{j\,\varphi_i(t)}$$

This will be developed in Part 3 and the corresponding example will be given in Part 4.

## 2-4. EMITTED SIGNALS AND PROCESSING

Let us remind that the optimal identification of $H(t,\varphi)$ is given by (3). When the autocorrelation function of the emitted signal is chosen as in (4) or approximatively, the identification is greatly simplified because the intercorrelation $\Gamma_{rs}$ gives the look of H directly.

Signals that have been studied and used in our laboratory for many years are PSK signals (Phase Shift Keying), i.e. carrier frequency $\nu_o$ is phase-modulated by a maximal length binary sequence $c(t) = (\pm 1)$

$$s(t) = c(t) \sin 2\pi\nu_o t \tag{6}$$

The two phasis states are 0 and $\pi$ .

### 2-4-1. Classical correlation properties of these signals.

The binary sequence c(t) is periodic, with period $T = N\theta$, where $\theta$ is the elementary time interval. Its autocorrelation function is periodic with the same period, and triangular as indicated in Figure 6. The maximum of $\Gamma_c(\tau)$ is N for $\tau = kT$, and the basis is -1. So, when $N \gg 1$, $\Gamma_c(\tau)$ looks like $N.\delta(\tau)$. The "performance" of c(t) is directly connected to the value of N.

PSK signal s(t) is illustrated in Fig. 6. The product $\theta.\nu_o$ is chosen to be an integer. The higher k is, the smaller the relative bandwidth of s(t) is. Autocorrelation function $\Gamma_s(\tau)$ is given in Fig. 6 too ; it is also periodic and modulated, and can be approximated over one period by

$$\Gamma_s(\tau) \sim \Gamma_s(0) . \Lambda_\theta(\tau) \cos 2\pi\nu_o\tau \tag{7}$$

where $\quad \Lambda_\theta(\tau) = 1 - \frac{|\tau|}{\theta} \quad , \quad |\tau| < \theta$

So these signals are a good choice for identification ; besides, the binary phasis modulation is easy to produce and process. A PSK generator with many choices of $\nu_o$ and N has been built and used in our laboratory for many years [10].

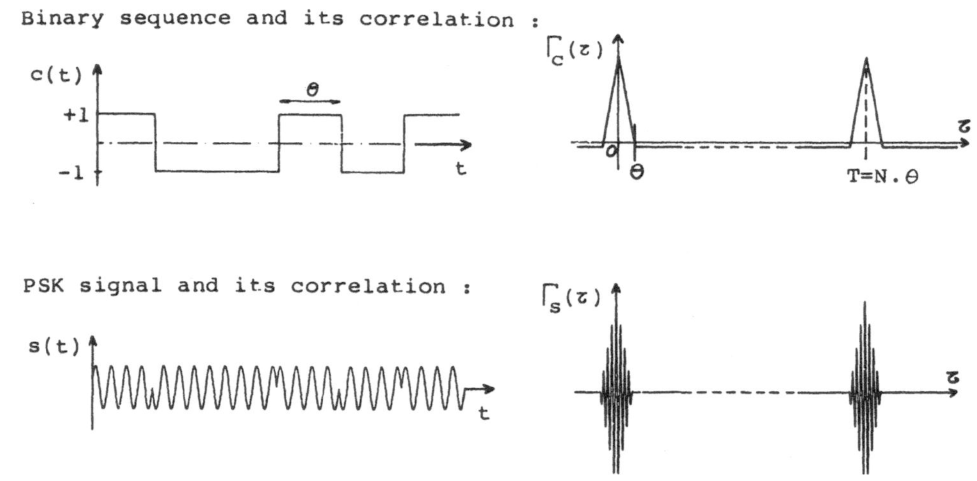


FIGURE 6 : Signals s(t) , c(t) and their autocorrelations

2-4-2. Processing.

It essentially consists in the intercorrelation of s(t) and r(t). This intercorrelation can be implemented directly. Yet, given the bandpass property of emitted and received signals, it is often simpler to correlate their low pass components themselves (*). In the general case, a 2x2 low pass component correlation must be evaluated [13,15] ; but s(t) given by (6) has only one low pass component which is c(t) ; so processing is usually operated according to the classical schema below (Fig. 7) : complex demodulation of received signal, intercorrelation of the two components with c(t) (this can be equivalently obtained by a matched filter whose i.r. is c(-t)). After that, different operations are possible and particularly envelope detection which leads to an estimation of $(H(\varphi))^2$.

---

(*) Let us recall that each bandpass signal s(t) round a carrier frequency $\nu_o$ can be written as :

$$s(t) = P(t)\cos 2\pi\nu_o t - Q(t)\sin 2\pi\nu_o t.$$

P and Q are the low pass components of s(t) . P(t) + j Q(t) is the complex amplitude of s(t) relative to $\nu_o$. This can be used as soon as the bandwidths of P and Q are inferior to $\nu_o$ ; in many practical applications, these bandwidths are very small comparatively to $\nu_o$.

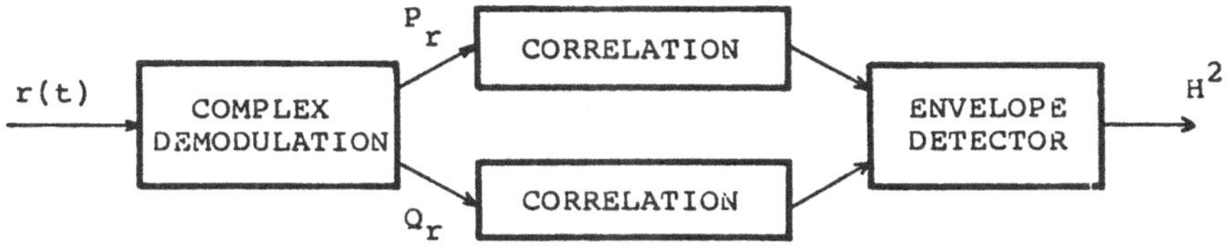


FIGURE 7 : Two component receive processing

As emission is periodic, the processing of Figure 7 is repeated every T. This processing is a stationary intercorrelation, so it assumes that the channel response does not vary during T. But, from one emission to another, it is possible to obtain successive estimations of $H(\varphi)$, and so to exhibit time variations of $H(\varphi)$ (slower than T).

NOTE : The estimation of more rapid evolution of $H(t, \varphi)$ with respect to t needs either a non stationary intercorrelation (like 3'), or a generalized intercorrelation (time frequency interambiguity [13]).

## 3. TIME DELAY ESTIMATION. PERFORMANCE

The intercorrelation (3) or (3'), using well chosen s(t), gives a second order optimal non parametric identification of $H(t,\varphi)$ whatever it looks. When $H(t,\varphi)$ can be a priori modelled by (4), channel identification can be proceeded by means of the estimation of amplitude, delay and phasis parameters related to different paths. In underwater acoustics, this model is effectively often adopted.

In a first step, we recall the basis of parameter estimation. Then, we study time delay estimation more precisely, and the interesting and new problem of jointly estimating time delay and phasis.

### 3-1. SOME ELEMENTS OF CLASSICAL PARAMETER ESTIMATION THEORY

Classical model [18,20] considers the received signal to be the addition of a known signal $y(t, \underline{\theta})$ which is a function of parameter $\underline{\theta}$ (vectorial since there may be n parameters), and noise b(t). This noise b(t) will always be assumed stationary, zero mean, and white (to simplify the calculation), with power spectral density (psd) $\gamma_o$. All signals are assumed real.

$$r(t) = y(t, \underline{\theta}) + b(t) \tag{8}$$

Let us insist on the point that the form of the signal y is assumed to be KNOWN here (opposite to the hypothesis of PASSIVE time delay estimation, which is another classical problem in sonar [14]).

Moreover, only the case when parameter $\underline{\theta}$ is NON RANDOM is studied here (which is coherent with the hypothesis of § 2).

For example, in the multipath case, $y(t,\underline{\theta})$ is

$$y(t,\underline{\theta}) = \sum_i \alpha_i \, s(t - \tau_i) \tag{9}$$

which corresponds to a real and non time varying i.r. $H(\varphi)$

$$H(t, \varphi) = H(\varphi) = \sum_i \alpha_i \, . \, \delta(\varphi - \tau_i)$$

By using the probability density function (pdf) of b(t) -it will be assumed gaussian- it is possible to use a stronger criterion than the second order one used in § 2-1. Maximum likelihood (ML) theory will be used in the following [20], which leads to an optimal strategy for parameter estimation and to the measure of performance of ML estimator.

With the above hypothesis, classical results are :

i) Structure of the optimal estimator with white gaussian noise (wgn).

Receiving (8), the optimal vectorial ML estimate $\underline{\theta}$ must satisfy the n following equations .

$$\int [r(t) - y(t, \underline{\theta})] \frac{\partial y(t, \underline{\theta})}{\partial \theta_i} dt \Big|_{\underline{\theta}=\hat{\underline{\theta}}} = 0 \qquad i = 1 \text{ to } n \tag{10}$$

Particularly, when the signal energy $\int | y(t,\tau) |^2 dt$ is independent of the parameter $\theta_i$, the corresponding $i^{th}$ equation can become simplified into

$$\int r(t) \frac{\partial y(t, \underline{\theta})}{\partial \theta_i} dt \Big|_{\underline{\theta}=\hat{\underline{\theta}}} = 0 \tag{11}$$

ii) Performance of the optimal estimate $\hat{\underline{\theta}}$ .

We are essentially interested in the second order properties of $\hat{\underline{\theta}}$ : bias $\underline{B}$ (vector), and covariance matrix $\underline{\Gamma}_{\hat{\underline{\theta}}}$ .

$$\underline{B} = E\{\hat{\underline{\theta}}\} - \underline{\theta} \quad ; \qquad \underline{\underline{\Gamma}}_{\hat{\theta}} = E\{ (\hat{\underline{\theta}} - E\{\hat{\underline{\theta}}\})( \hat{\underline{\theta}} - E\{\hat{\underline{\theta}}\} )^T \qquad (12)$$

The estimate $\hat{\underline{\theta}}$ is non biased if $\underline{B} = 0$.

The dispersion is minimum when the whole information, contained in the observed signal, about $\underline{\theta}$ , is used. This information quantity is characterized by the "Fisher information matrix" :

$\underline{\underline{I}}(\underline{\theta}) = [I_{ij}]_{n \times n}$ , which terms $I_{ij}$ are given here by

$$I_{ij} = \frac{1}{\gamma_0} \int \frac{\partial y(t,\underline{\theta})}{\partial \theta_i} \cdot \frac{\partial y(t,\underline{\theta})}{\partial \theta_j} dt \qquad (13)$$

The estimate is said efficient if the dispersion reaches its minimal bound, called Cramer Rao bound, given in the case of a non biased estimate, by

$$\boxed{\text{CR bound} = \underline{\underline{I}}(\underline{\theta})^{-1}} \qquad (14)$$

## 3-2. TIME DELAY ESTIMATION

Let us apply the above results when estimating only one unknown time delay, noted $\tau$ , which is assumed non random, real and constant. (8) becomes

$$r(t) = \alpha\, s(t - \tau) + b(t) \qquad (15)$$

where $\alpha$ and s are known. b(t) is a wgn.

### 3-2-1. Optimal structure of $\hat{\tau}$

In this case, signal energy $\int | y(t,\tau)|^2 dt$ is independent of $\tau$.

So we use (11) to get the optimal estimator $\hat{\tau}$ of $\tau$ :

$$\int r(t) \frac{\partial s(t-\tau)}{\partial \tau} dt \Big|_{\tau=\hat{\tau}} = 0 \qquad (16)$$

The optimal structure is given in Figure 8a.

Practically, the two operations, derivation and integration, of (16) can be inverted. So, in a first time, r(t) is cross-correlated with s(t, $\tau$), and the optimal $\hat{\tau}$ is obtained by maximizing this intercorrelation. The new optimal schema becomes Fig. 8b. So this corresponds exactly to the intercorrelation structure used for non parametric identification of § 2. Here, the only value of optimal $\hat{\tau}$ is of interest.

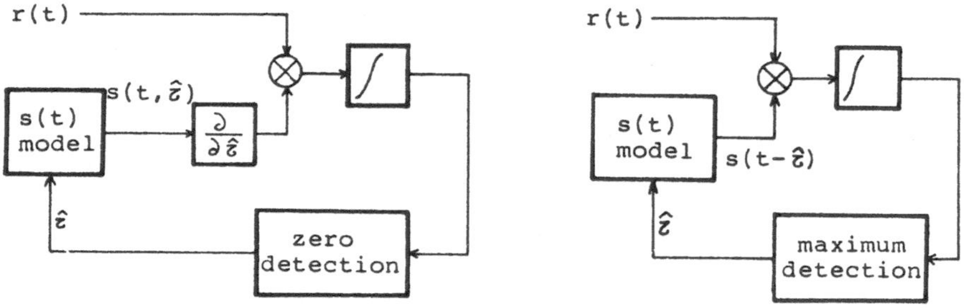


fig.8a fig.8b

FIGURE 8 : optimal estimator of $\tau$

3-2-2. Performance

Let R be the signal to noise ratio :

$$R = \frac{\alpha^2 \Gamma_s(0)}{\gamma_o} \tag{17}$$

$\Gamma_s(o) = \int s^2(t)\, dt$ is the energy of s(t).

Using some classic results, the optimal solution $\hat{\tau}$ of (16) is

i) asymptotically unbiased ($R \longrightarrow \infty$)
ii) asymptotically efficient

The Cramer Rao bound is given by

$$\boxed{\sigma_{\hat{\tau}}^{2} \geqslant \frac{1}{R\, B_s^2}} \tag{18}$$

where $B_s$ is the effective bandwidth of s(t) ; if $S(\nu) \rightleftharpoons s(t)$

$$B_s^{\,2} = \frac{4\pi^2}{\Gamma_s(0)} \int \nu^2 \, | S(\nu) |^2 \, d\nu = - \frac{\ddot{\Gamma}_s(0)}{\Gamma_s(0)} \tag{19}$$

Let us apply these relations when using the signal described in Section 2.

a) When low frequency transmission is performed, i.e. when the emitted signal is the binary sequence c(t) defined in 2-3, the psd $|S(\nu)|^2$ is a $\mathrm{sinc}^2 x$ centered on the $\nu$-axis origin. The integral in (19) is convergent only if s(t) is low pass filtered : for example we will choose to cut off all the frequencies beyond n sidelobes of $\mathrm{sinc}^2$.

Let $B_c$ be the obtained effective bandwidth and $E_c$ the sequence energy.

b) When bandpass transmission is performed, the emitted signals are PSK signals round $\nu_o$ (cf (6)) :

$$s(t) = c(t)\sin 2\pi\nu_o t \;,\; c(t) \rightleftharpoons C(\nu).$$

Let us calculate the effective bandwidth $B_s$ of s(t) :

$$|S(\nu)|^2 = \frac{1}{4}\left[\,|C(\nu-\nu_o)|^2 + |C(\nu+\nu_o)|^2\,\right] \qquad (20)$$

We also assume that only n sidelobes of $\mathrm{sinc}^2$ round $\nu_o$ are used (bandpass hypothesis). (19) becomes

$$B_s^2 = \frac{4\pi^2}{E_s}\int (2\nu^2 + 2\nu_o^2)\,|C(\nu)|^2\,d\nu = \frac{E_c B_c^2 + 4\pi^2\nu_o^2 E_c}{2\,E_s}$$

with (6) $\qquad E_c = 2\,E_s \qquad (21)$

$$\boxed{B_s^2 = B_c^2 + 4\pi^2\nu_o^2} \qquad (22)$$

So it clearly appears that in order to obtain the minimal variance $\sigma_{\hat{\varepsilon}}^2$ , one must use the maximal $B_s$ : and (22) leads to choose emitted signals whose carrier frequency $\nu_o$ is the highest ; particularly PSK signals are more performant than low frequency sequences c(t).

NOTE : this is valid only if the model (15) corresponds exactly to the received signal, i.e. if the channel acts only as a delay line whatever the transmitted signal may be.

c) Validity of this model in underwater acoustics

i) In fact, as we have seen before, the undersea channel is represented by MULTIPLE time delays (cf Figure 5). If these delays are different enough from one another, they can be estimated separately. This corresponds to the following treated example. Lately, we have been studying the case of close delays.

ii) But, as it has already been mentioned, the model (15) is valid in underwater acoustics, from our experience, only in a case of low frequency transmission. When transmitted signals are bandpass round the carrier $\nu_o$, the channel propagation includes, in addition, a PHASE SHIFT on the carrier frequency $\nu_o$.

So we are now to study this new situation which leads to new results.

3-3. JOINT ESTIMATION OF TIME DELAY AND PHASIS.

The emitted signal is still (6) :

$$s(t) = c(t) \sin 2\pi \nu_o t$$

and the channel introduces a time delay $\tau$ and a phase shift $\varphi$ so that the received signal is

$$r(t) = \alpha\, c(t-\tau) \sin[2\pi \nu_o(t-\tau)+ \varphi] + b(t) = \alpha\, s(t,\tau,\varphi) + b(t) \quad (23)$$

The other quantities are the same as before. The question of interest is now the joint estimation of $\tau$ and $\varphi$ .

3-3-1. Structure

In this case too, the norm of signal $s(t,\tau,\varphi)$ is independent of both parameters to estimate. Thus (11) is still used for optimal joint ML estimation. The optimal estimates $\hat{\tau}$ and $\hat{\varphi}$ satisfy the two following equations :

$$\left. \int r(t) \frac{\partial s(t,\tau,\varphi)}{\partial \tau} dt \right|_{\tau=\hat{\tau},\varphi=\hat{\varphi}} = 0$$

$$\left. \int r(t) \frac{\partial s(t,\tau,\varphi)}{\partial \varphi} dt \right|_{\tau=\hat{\tau},\varphi=\hat{\varphi}} = 0 \quad (24)$$

Still assuming that c(t) is filtered in the same way as above, s(t) is bandpass and its quadrature signal can be written as

$$s_Q(t) = -c(t)\cos 2\pi \nu_0 t \tag{25}$$

By developing $s(t,\tau,\varphi)$ as

$$s(t,\tau,\varphi) = \cos\varphi \,.\, s(t-\tau) - \sin\varphi \,.\, s_Q(t-\tau) \tag{26}$$

then the first member of the first equation of (24) for example becomes

$$\int r(t)\,\frac{\partial\, s(t,\tau,\varphi)}{\partial \tau}\,dt = \cos\varphi \,.\, \frac{\partial}{\partial \tau}\left\{\Gamma_{rs}(\tau)\right\} - \sin\varphi \,.\, \frac{\partial}{\partial \tau}\left\{\Gamma_{rs_Q}(\tau)\right\}$$

where $\Gamma_{rs}$ and $\Gamma_{rs_Q}$ are obvious intercorrelations. The first equation of (24) leads to

$$\operatorname{tg}\hat{\varphi} = + \frac{\dot{\Gamma}_{rs}(\hat{\tau})}{\dot{\Gamma}_{rs_Q}(\hat{\tau})} \tag{27}$$

The second equation of (24) is developed with the same expressions and leads to :

$$\operatorname{tg}\hat{\varphi} = - \frac{\Gamma_{rs_Q}(\hat{\tau})}{\Gamma_{rs}(\hat{\tau})} \tag{28}$$

The optimal receiver for the joint estimation of $\tau$ and $\varphi$ is elaborated using (27) and (28) and is figured below :

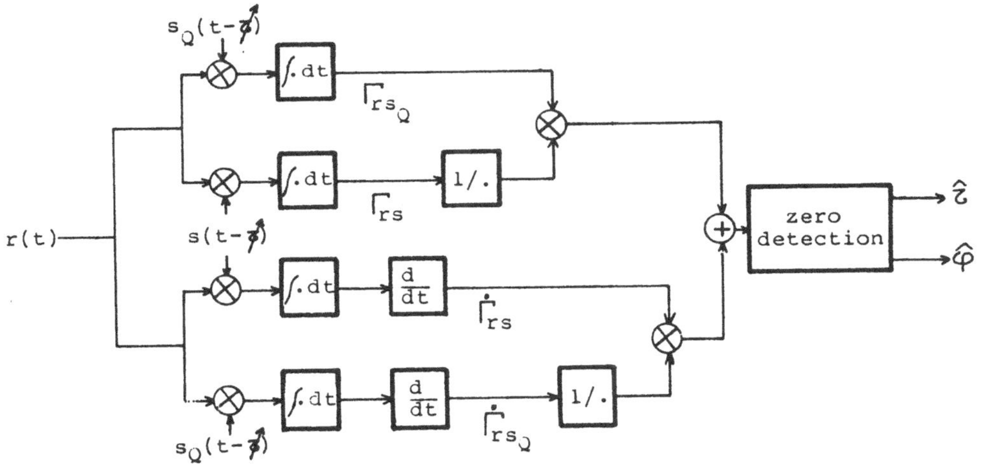


FIGURE 9 : Theoretical optimal joint estimator of $\tau$ and $\varphi$ .

Such an implementation is not very easy, especially because of the starting derivation. Another implementation equivalent to (27) + (28) is looked for, noting that only the intercorrelations $\Gamma_{rs}$ and $\Gamma_{rs_Q}$ and their derivatives are used. By writing (27) = (28)

$$\operatorname{tg}\hat{\varphi} = + \frac{\dot{\Gamma}_{rs}(\hat{\tau})}{\dot{\Gamma}_{rs_Q}(\hat{\tau})} \qquad = - \frac{\Gamma_{rs_Q}(\hat{\tau})}{\Gamma_{rs}(\hat{\tau})}$$

$$\dot{\Gamma}_{rs}(\hat{\tau}).\ \Gamma_{rs}(\hat{\tau}) + \dot{\Gamma}_{rs_Q}(\hat{\tau}).\ \Gamma_{rs_Q}(\hat{\tau}) = 0 \qquad (29)$$

The first term of (29) represents the derivative with respect to $\tau$ of a term like $\Gamma^2_{rs}(\tau) + \Gamma^2_{rs_Q}(\tau)$. Besides, $\hat{\varphi}$ is obtained by (28)

$$\hat{\varphi} = - \operatorname{Arctg}\left( \Gamma_{rs_Q}(\hat{\tau}) \ / \ \Gamma_{rs}(\hat{\tau}) \right) \qquad (29')$$

Thus a new optimal receiver is now given by the Figure 10. The received signal r(t) is intercorrelated with the replicas of s(t) and $s_Q(t)$ ; then both intercorrelations are on one hand squared and summed, and on the other hand divided.

Optimal $\hat{\tau}$ is the abscissa of the maximum of square summation (or its root : envelope), and using this optimal value $\hat{\tau}$ in the quotient (29') we get optimal $\hat{\varphi}$ .

Let us note that the first branch of this schema (which leads to $\hat{\tau}$ ) is the same as the processing used in § 2-4 for non parametric identification (reception with two quadrature components, and envelope detection). Here, the calculation of Arctg gives the phasis $\hat{\varphi}$ too.

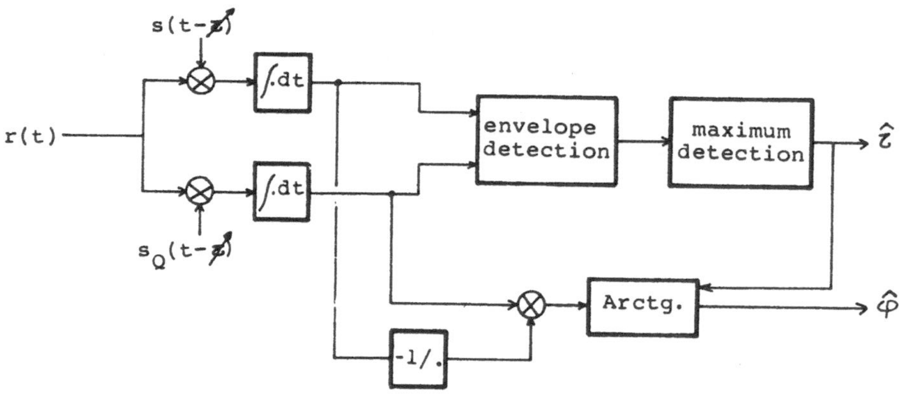


FIGURE 10 : Another optimal estimator of $\tau$ and $\varphi$

3-3-2. Performances

Let us calculate the 4 terms of the 2x2 Fisher matrix given by (13) :

$$I_1 = \frac{\alpha^2}{\gamma_o} \int \left[ \frac{\partial s(t, \tau, \varphi)}{\partial \tau} \right]^2 dt$$

and using (26)

$$I_1 = \frac{\alpha^2}{\gamma_o} \int \left[ \frac{\partial s(t-\tau)}{\partial \tau} \cos\varphi - \frac{\partial s_Q(t-\tau)}{\partial \tau} \sin\varphi \right]^2 dt$$

$$= \frac{\alpha^2}{\gamma_o} [ \cos^2\varphi . \Gamma_{\dot{s}}(0) + \sin^2\varphi . \Gamma_{\dot{s}_Q}(0) - \sin 2\varphi . \Gamma_{\dot{s}\dot{s}_Q}(0) ]$$

But $\Gamma_{\dot{s}}(0) = \Gamma_{\dot{s}_Q}(0) = - \ddot{\Gamma}_s(0)$

$$\Gamma_{ss_Q}(0) = 0 = \Gamma_{ss}(0) \qquad (30)$$

Using (19) and (17), we obtain finally

$$\boxed{I_1 = R\, B_s^2} \qquad (31)$$

$$I_2 = \frac{\alpha^2}{\gamma_o} \int \left[ \frac{\partial s(t, \tau, \varphi)}{\partial \varphi} \right]^2 dt$$

Using the same development, and noting that

$$\Gamma_{s_Q}(0) = \Gamma_s(0) = E_s \quad ; \quad \Gamma_{ss_Q}(0) = 0 \qquad (32)$$

we obtain

$$\boxed{I_2 = R} \qquad (33)$$

$$I_{12} = I_{21} = \frac{\alpha^2}{\gamma_o} \int \frac{\partial s(t, \tau, \varphi)}{\partial \tau} \cdot \frac{\partial s(t, \tau, \varphi)}{\partial \varphi} dt$$

$$= \frac{\alpha^2}{\gamma_o} [\cos\varphi . \sin\varphi . \Gamma_{\dot{s}s}(0) + \sin^2\varphi . \Gamma_{\dot{s}_Q s}(0) - \cos^2\varphi . \Gamma_{\dot{s}s_Q}(0) - \sin\varphi . \cos\varphi . \Gamma_{\dot{s}_Q s_Q}(0)]$$

The two extrema terms are zero (cf (30)) ; besides $\Gamma_{\dot{s}_Q s}(0) = - \Gamma_{\dot{s}s_Q}0)$

So $$\boxed{I_{12} = I_{21} = \frac{\alpha^2}{\gamma_o} \cdot \dot{\Gamma}_{s_Q s}(0)} \qquad (34)$$

First, (34) shows an important result : the estimates $\hat{\tau}$ and $\hat{\varphi}$ are always CORRELATED. Let us calculate the quantity $\dot{\Gamma}_{s_Q s}(0)$ which measures this coupling.

By using relationships between s and $s_Q$, we obtain

$$\Gamma_{\dot{s}_Q s}(0) = 4\pi \int \nu \, | S(\nu) |^2 \, d\nu$$

But $| S(\nu) |^2$ is given by (20) ; so

$$\Gamma_{s_Q s}(0) = \pi \nu_0 \, \Gamma_c(0) = \pi \nu_0 E_c = 2\pi \nu_0 E_s \qquad (35)$$

The bound of each estimate is calculated with (14).

The determinant of I is

$$\Delta = I_1 \, I_2 - I_{12}^2 = R^2 \, [B_s^2 - 4\pi^2 \nu_0^2] = R^2 \, B_c^2$$

So, the inferior bound of Var $\hat{\tau}$ is $I_2/\Delta$

$$\boxed{\text{Var } \hat{\tau} \geqslant \frac{1}{R \, B_c^2}} \qquad (36)$$

This is a very important result : if $\varphi$ were perfectly known, only the time delay $\tau$ would be to estimate (case of § 3.2) ; and the effective bandwidth which would occur, would be $B_s >> B_c$ . When a carrier phasis shift has also to be estimated, the quality of the estimate $\hat{\tau}$ is connected to the only low-frequency part of s(t), that is to say to the effective bandwidth of the sequence c(t).

The Inferior bound of var $\hat{\varphi}$ is now

$$\boxed{\text{Var } \hat{\varphi} \geqslant \frac{I_1}{\Delta} = \frac{1}{R} \, \frac{B_s}{B_c} = \text{Var} \, \hat{\tau}_{min} \cdot B_s^2} \qquad (37)$$

It turns out that $\varphi$ will be poorly estimated.

## 4. AN EXAMPLE IN UNDERWATER ACOUSTICS

For many years, the CEPHAG has been carrying out undersea experiments. These were made in collaboration with the GERDSM (Groupe d'Etude et Recherche en Détection Sous-Marine, Le Brusc, France). The methodology developed in § 2 has been implemented for various undersea propagation cases, in various conditions : horizontal or vertical propagation, long or short ranges, different emitter and receiver dippings, fixed or mobil emitter and receiver, different locations, various emitter frequencies,... The example given in Figure 3 corresponds to a horizontal long distance propagation case.

We are to detail now an experiment which is currently studied ; it will permit us to illustrate the above developed concepts about active identification and time delay estimation.

## 4-1. DESCRIPTION OF THE EXPERIMENT

(it was conducted in collaboration with the GERDSM)

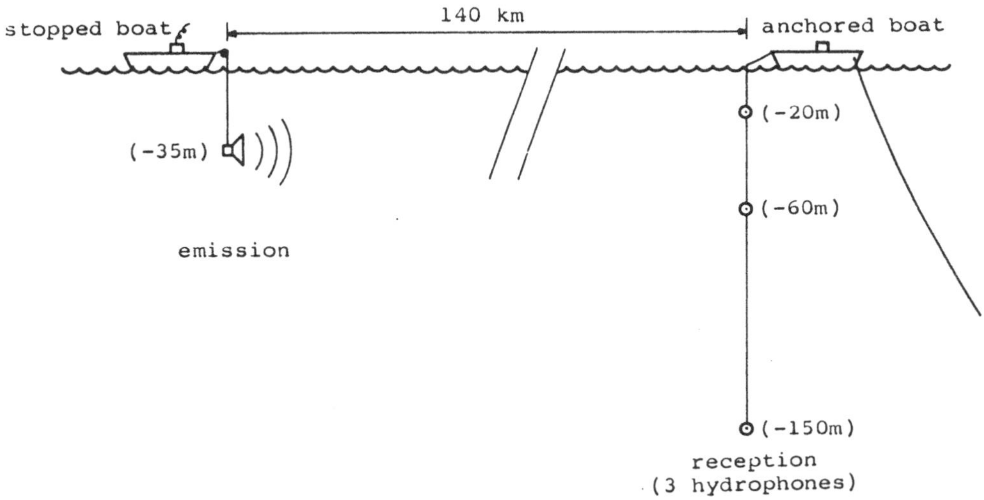


FIGURE 11 : Description of the experiment

A transmission channel is defined between the emitter and each of the three receivers. Geometric conditions are precised on the schema. The emitter boat is stopped. The receiver boat is at anchor.

Emitted signals are those defined in (6), with a carrier frequency $\nu_o = 60$ Hz. The binary sequence used for modulating $\nu_o$ is characterized by N = 127 and the elementary time interval is $\theta = 67$ ms. Then signal correlation function is sharp enough to identify the channel impulse response correctly.

Emission is periodic, with period $T = 127.\theta \sim 8.5$ sec, during a total time interval of about 2 H.

## 4-2. PRIMARY RESULTS OF NON PARAMETRIC IDENTIFICATION

a) Using the method developed in § 2 -intercorrelation between emitted signal at E and each of the three received signals- the following plot is obtained for one period of the emitted signal.

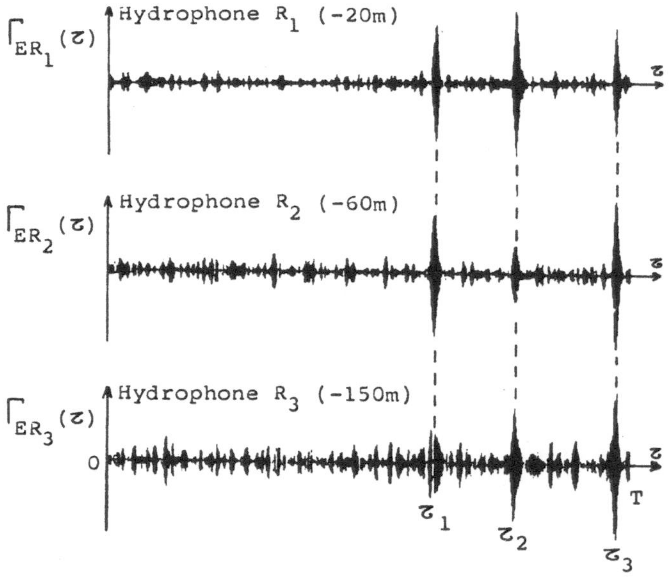


FIGURE 12 : Intercorrelation E - $R_k$ , k = 1 to 3, for one period T of the emitted signal.

The principal characteristic aspects of the transmission appear in Figure 12.

i) first, 3 main paths are exhibited through each of the 3 channels, with slightly different amplitudes.

ii) the third channel E - $R_3$ is clearly more noisy than the two others.

Let us note that, obviously, all that appears outside the three "peaks" may be additive noise or secondary peaks as well.

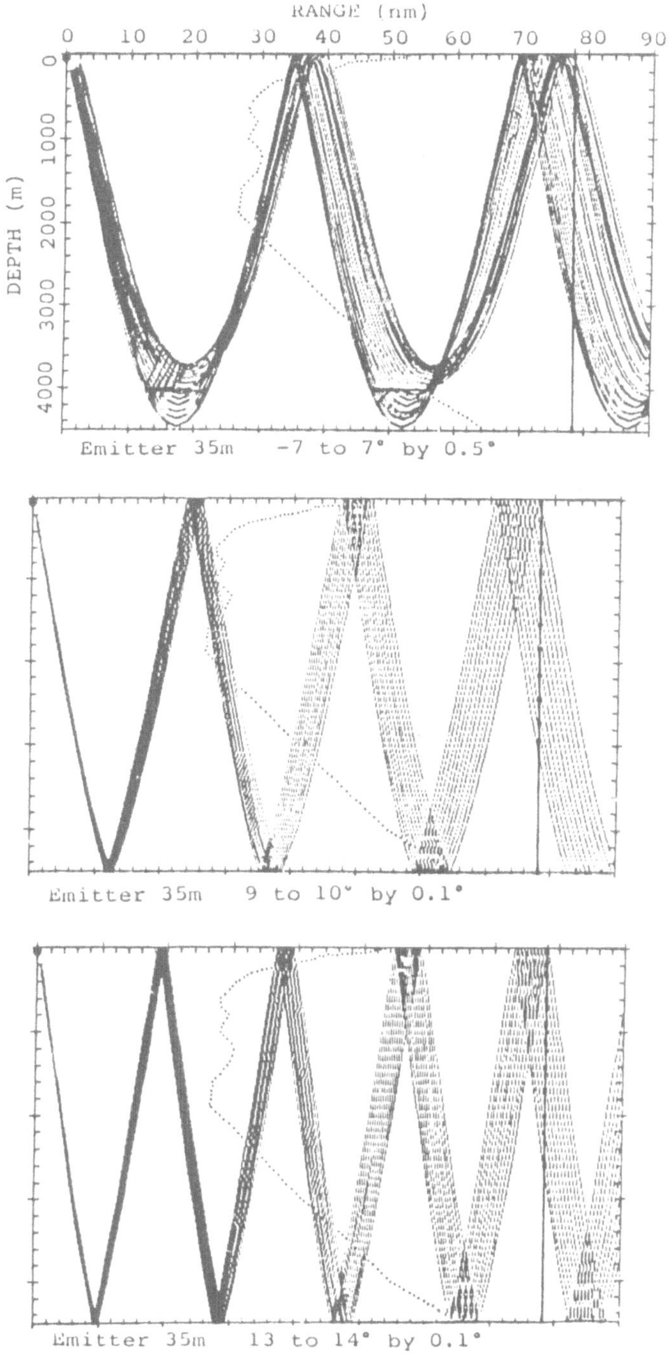


FIGURE 13: Ray-tracing corresponding to the experiment of Fig.11 (from GERDSM)

b) The ray tracing corresponding to this propagation situation takes account of emitter and receiver geometry, bathymetry and corresponding celerity profile (Fig. 13). Time delay deducted from Figure 13 exhibits three main delays from three more energetic rays, as shown in Figure 14.

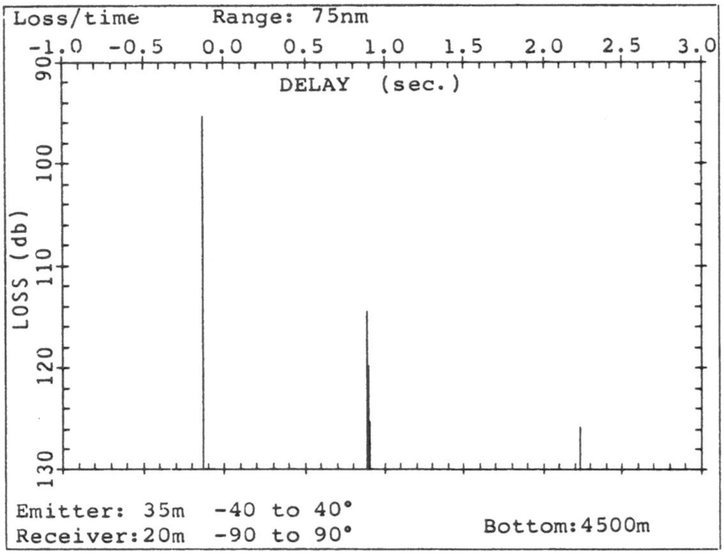


FIGURE 14 : Prediction of delays by ray-tracing method on receiver $R_1$. (from GERDSM).

This example exhibits the complementarity of both approaches : approxi-mative time delays are given in Figure 14 by ray-tracing, whereas the channel time evolution will be followed during all emission duration, as we shall see later.

The following developments concern only the first channel E - $R_1$ for which the signal to noise ratio (SNR) is higher. It has been estimated at about

$$R \sim 11 \text{ dB} \tag{38}$$

## 4-3. TEMPORAL EVOLUTION OF THE CHANNEL IMPULSE RESPONSE (first hydrophone)

We first calculate the envelope of $E - R_1$ intercorrelation (or matched filter).

The figure 15 gives the temporal evolution of this envelope during about 1/2 H.

So it turns out that the channel $E - R_1$ can be effectively modelled as being the sum of 3 delays, the amplitudes being quasi constant. This can a posteriori justify the deterministic model chosen for $H(\varphi)$.

$$H(\varphi) = \sum_i \alpha_i \cdot \delta(\varphi - \tau_i)$$

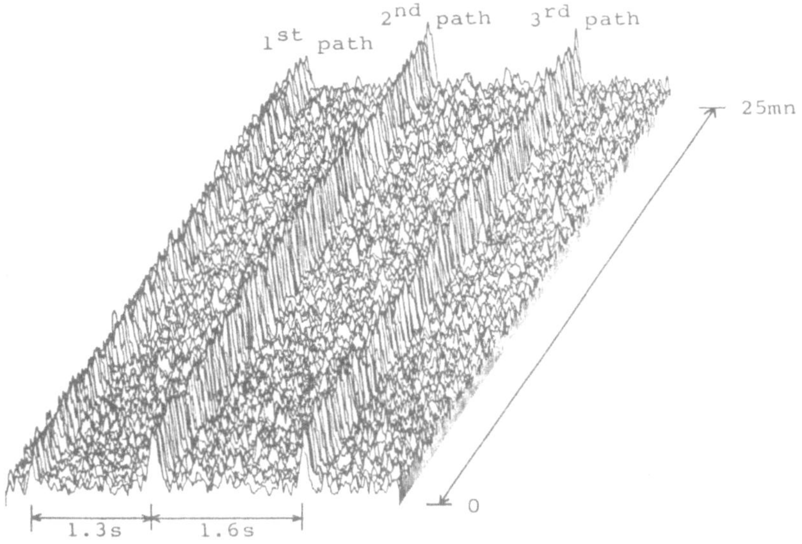


FIGURE 15 : Temporal evolution of the i.r. envelope of $E$-$R_1$ channel.

A first estimation of the delays between the three main paths, given in figure 15, points out the 1.3 and 1.6 sec. values, which are significantly different from those given by the ray-tracing method, 1.0 and 1.3 sec. respectively.

Using many successive emissions, then as many successive received signal realizations, $r(t) = \sum_i \alpha_i \, s(t - \tau_i) + b(t)$, the above method (§ 3) can be applied in order to ML estimate the 3 delays $\tau_i$.

## 4-4. TIME DELAYS ML ESTIMATION

The different delays being sufficiently different from one another in the previous figure, it is clear that each of them may be processed as if it were alone. This has been implicitly achieved by intercorrelating r(t) with a copy of emitted signal.

a) Choice of the model and optimal estimator : let us detail one of the three peak patterns of the Figure 12.

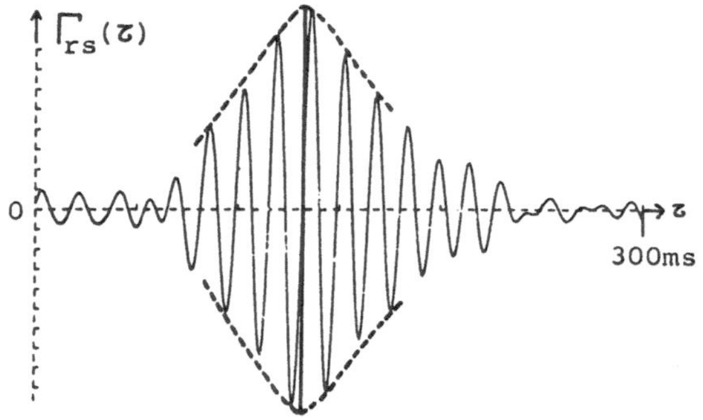


FIGURE 16 : Zoom on a peak pattern of Figure 12.

It turns out that the abscissa of the envelope maximum is DIFFERENT from that of signal intercorrelation. This means that a transmission model leading to an only time delay is not sufficient, and that a phasis shift must be introduced. Effectively, if the received signal is modelled by (23) for each delay, by intercorrelating r(t) with the emitted signal (6) we obtain

$$\Gamma_{rs}(\theta) = \int r(t)\ s(t-\theta)\ dt$$

$$= \propto\ [c(t-\tau)\sin(2\pi\nu_0(t-\tau)+\varphi)+b(t)][c(t-\theta)\sin 2\pi\nu_0(t-\theta)dt]$$

The second term of the integral is about 0, and the first one becomes

$$\Gamma_{rs}(\theta) \neq \frac{1}{2} \int c(t-\tau)\ c(t-\theta)\ .\ \cos(2\pi\nu_0(\theta-\tau)+\varphi)\ dt$$

$$\Gamma_{rs}(\theta) = \frac{1}{2} \cos\ [2\pi\nu_0(\theta-\tau)+\varphi]\ .\ \Gamma_c(\tau-\theta) \qquad (39)$$

which is not maximum for $\tau = \theta$ (value which maximizes the envelope $\Gamma_c(\tau-\theta)$ ) because of $\varphi$ .

SO THE MODEL WE HAVE TO CONSIDER CORRESPONDS TO (23). THE OPTIMAL ESTIMATE OF $\tau$ AND $\varphi$ must be elaborated according to figure 13 and $\tau$ is given by the abscissa of the envelope maxima.

The following figure gives the evolution of $\hat{\tau}_i$ , i = 1 to 3.

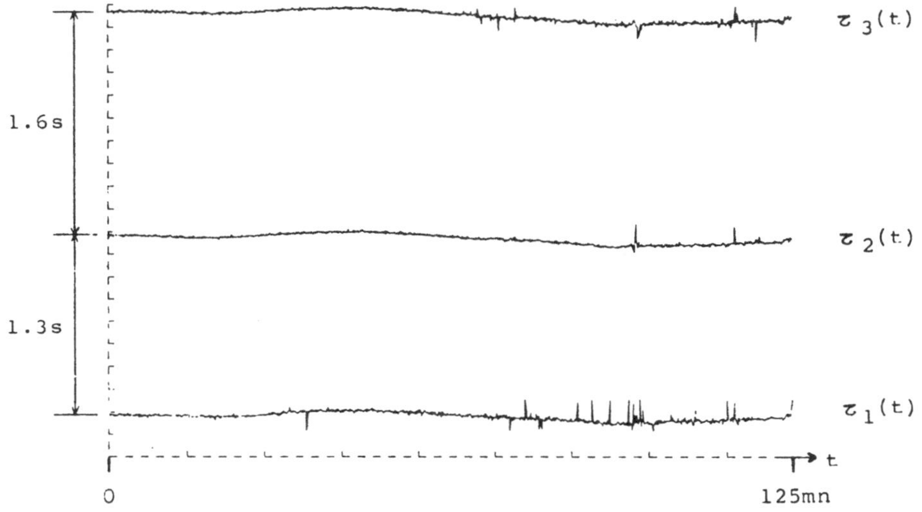


FIGURE 17 : Evolution of $\hat{\tau}_i$.

The following statistical results (calculated on about 450 realizations) are obtained :

| $\hat{\tau}_i$ | E {·} (from arbitrary origin) | $\sigma_{\hat{\tau}}$ | $\hat{\Delta}\tau$ | E{$\hat{\Delta}$} | $\sigma_{\hat{\Delta}}$ |
|---|---|---|---|---|---|
| $\hat{\tau}_1$ | 0.295 s | 32.9 ms | $\hat{\tau}_2 - \hat{\tau}_1$ | 1.296 s | 24.3 ms |
| $\hat{\tau}_2$ | 1.591 s | 33.3 ms | $\hat{\tau}_3 - \hat{\tau}_2$ | 1.612 s | 18.1 ms |
| $\hat{\tau}_3$ | 3.202 s | 39.9 ms | $\hat{\tau}_3 - \hat{\tau}_1$ | 2.908 s | 29.3 ms |

We notice that $\mathrm{Var}\,\hat{\Delta}_i = \sigma_{\Delta_i}^2$ is inferior to the sum of the varian-ces of the corresponding $\hat{\tau}_i$ ; that means that there exists a correlation between $\hat{\tau}_i$. In fact, a part of this correlation at least is obvious in figure 17 : slow variations are present in the 3 $\hat{\tau}_i$ estimates. This probably corresponds to equipment movements, and is not directly related to noise or channel. By cancelling these very low frequency components of $\hat{\tau}_i$ (inferior to 10 mn), we obtain now

| $\hat{\tau}_i$ | $\sigma_{\hat{\tau}}$ |
|---|---|
| $\tau_1$ | 10 ms |
| $\tau_2$ | 6.5 ms |
| $\tau_3$ | 11 ms |

b) Comparison with expected results.

b1) Let us calculate the Cramer Rao bound of $\mathrm{Var}\,\hat{\tau}_i$. Using the signal characteristics given § 2-1, $B_S$ and $B_C$ are calculated by no-ting that the bandwidth of the filtered signal is $\pm$ 30 Hz round $\nu_o$ (this corresponds effectively to hydrophone filtering), then

$$B_S \sim 380 \text{ Hz} \quad ; \quad B_C \sim 47 \text{ Hz} \tag{40}$$

Then the Cramer Rao bound given by (36) becomes with (40) and (38)

$$\sigma_{\hat{\tau}_i} \geqslant 5.6 \text{ ms} \tag{41}$$

Then the results we obtain are quite comparable with this performance ; the last variance calculation of the above table is in very good agreement with this.

b2) Performance of $\hat{\varphi}$ estimate

Let us now calculate the Cramer-Rao bound given by (37). We ob-tain $\sigma_{\hat{\varphi}} > 122°$, which is effectively a poor estimation of $\varphi$, and we cannot expect to estimate the phase shift in that way.

## 5. CONCLUSION

The optimal method (in the mean square sense) of the active non parametric identification of a linear channel has been recalled and developed here : it uses large bandwidth signals and input-output intercorrelation processing. When the channel is a multipath channel, this identification scheme has been shown to be still valid, and it enables us to optimally estimate -in maximum likelihood sense, now- time delays. The importance of SNR and effective bandwidth has been exhibited. Maximum length binary sequences, and corresponding PSK signals prove very suitable and performant for this identification. Furthermore, they are used for example in acoustic tomography [6], and many PSK signal generators are developed nowadays.

By applying these methods to undersea acoustics, different propagation acoustic paths can be studied. Time delays are well estimated and followed, and they have often been found rather stable during several minutes. It is possible to choose binary emitted signals in order to obtain a desired precision on path estimation. But one must take account, for each delay, of an additional carrier frequency phasis shift. This cannot be estimated sufficiently precisely in our case. But other studies [2] have shown that phasis differences between the different paths -not measured on the signal itself, but on the corresponding complex amplitudes- can be well estimated and can also remain stable during at least one minute : this enables us to envisage "path coherent equalization" for communications.

As regards to theoretical parameter estimation in multipath model, a lot of work is still to be done, in order to take account of all the elements (amplitudes, phasis) of the different cases ; and particularly the problem of very close paths has still to be studied.

**AKNOWLEDGEMENTS**

This research was supported by the Direction of the French Naval Constructions, under contract n° C 85.76.214.00.

**BIBLIOGRAPHY**


1. Bello P., Characterization of randomly time variant linear channel IEEE Trans. on Comm. Syst., Vol. CS11, 360-393, Dec. 1963.

2. Boucard H., G. Jourdain and G. Loubet, Traitement optimal linéaire de signaux de communication sous-marine expérimentale dans un canal à deux trajets de propagation, Coll. GRETSI, Nice, May 1985.

3. Chavand F. and M. Lechevallier, Transmission numérique à travers un canal ionosphérique simulé, Revue du CETHEDEC, 4e Trim. 1979, NS 79-2.

4. Collin C. and A. Marguinaud, Evaluation expérimentale de la sélectivité d'une liaison par diffusion troposphérique, Revue Technique Thomson CSF, Vol. 11, n° 1, March 1979.

5. Cunningham A.B., Some alternate vibrator signals, Geophysics, Vol. 44, n° 12, Dec. 1979.

6. Desaubies Y. and P. Tillier, Projet de tomographie acoustique de l'océan, Rapport IFREMER, Dec. 1984.

7. Ellinthorpe and Nuttal, Theoretical and empirical results on the characterization of undersea acoustic channel, First Annual Communication, IEEE Conf., Boulder, Juin 1965.

8. Faure B., J.Y. Jourdain, Estimation de la fonction de diffusion d'un milieu réel. Rapport CEPHAG n° 24/79.

9. Faure P. and M. Depeyrot, Eléments d'automatique, Dunod, 1974.

10. Henrioux J.P., Génération de séquences binaires, Rapport CEPHAG n° 5/72.

11. Jourdain G. and J.Y. Jourdain, Characterization of the underwater channel. Application to communication. Issues in Acoustic Signal, Image Processing and Recognition, NATO ASI Series, Edited by Chen, Springer Verlag, 1983.

12. Jourdain G. and G. Tziritas, Communication over fading dispersive channels, Optimal receivers and signals, Signal Processing, Vol. 6, Janvier 1984.

13. Jourdain G., Filtres linéaires aléatoires et non stationnaires, Thèse D.E., USMG/INPG, Grenoble, 1976.

14. Jourdain J.Y., Démodulation complexe. Rapport CEPHAG 1/80.

15. Knapp C.H. and G.C. Carter, The generalized correlation method for estimation of time delay, IEEE Trans. Acoust. Speech and Signal Proc., ASSP 24, n° 4, 320-327, 1976.

16. Lambert P., Contribution par l'étude de la phase transmise à la modélisation du milieu acoustique sous-marin, Thèse de Docteur-Ingénieur, INPG, Grenoble, March 1983.

17. Laval R., Caractérisation déterministe et stochastique du canal de transmission acoustique sous-marin, Coll. GRETSI, Nice, 1975.

18. Ocean Tomography Group, A demonstration of ocean acoustic tomography, Nature, Vol. 299, 121-125, Sept. 1982. (Behringer, Birdsall, Brown, Cornuelle, Heinmiller, Knox, Metzger, Munk, Spiesberger, Spindel, Webb, Worcester, Wunsch).

19. Sage A.P. and Melsa J., Estimation theory with applications to communications and control, Mac Graw Hill, 1971.

20. Turin G.L., Introduction to spread spectrum antimultipath techniques and their application to urban digital radio. Proc. of IEEE, Vol. 68, n° 3, 321-353, March 1980.

21. Van Trees H., Detection, estimation and modulation theory, Part I, J. Wiley, N.Y., 1971.

CHAPTER 3

# OPTIMAL FILTERING IN THE PRESENCE OF MULTIPATH

José M. F. Moura
Maria João D. Rendas

## 1. INTRODUCTION

The need for passively locating an emitter arises in many practical situations as, for example, in military surveillance, fish detection, or help systems. Eventhough the nature of the signals involved, and of the particular media they propagate in, are specific to each application, common features can be recognized: all these systems make use of a model of the radiated signal, and of the channel's action on it, and try to infer the emitter location from the estimated channel transfer function.

This work assesses the source position estimation problem in multipath ambient, i.e., when a discrete number of paths exist from source to receiver. Although the study is restricted to the case of acoustic waves propagating in the deep ocean, the multipath type of structure is characteristic of other media as well, e.g., in the propagation of electromagnetic waves in the atmosphere. For this kind of channels, the signal received at a point remote from the source consists of several attenuated and delayed versions of the radiated signal. This may result from reflecting boundaries and/or from a nonzero gradient of the propagation speed.

In part 2, issues related to channel modelling are considered. An algorithm is presented which predicts the number of distorted received replicas, and the delay and attenuation of each one for a multilinear velocity profile. For the bilinear velocity profile, the problem is reduced to the rooting of a 4th degree (trig nometric) polynomial. The result is generalized to a multilinear profile.

In part 3, the positioning problem is considered. Firstly, the maximum likelihood estimator of the source position for this type of channel modulation and under the hypothesis of Gaussian radiated signal and observation noise is derived. It is concluded that the estimator's structure is closely related to the assumed channel model. Secondly, two general signal classes are considered, and the receiver for each one presented. In the first one, the emitted signal is represented as a linear combination of a finite number of known basis functions. Several block decompositions for the receiver are presented, each one providing intuitively appealing interpretations of its structure. In essence, the estimator mimicks the assumed channel model, acomplishing a Maximum A Posteriori estimation of the (expansion) coefficients of the radiated signal. In the second one the signal is described as the output of a finite dimensional linear system driven by white noise. The filter is designed using Kalman-Bucy type techniques.

At the end of the paper, before the references, we help the reader with a short partial table of symbols with their meaning.

## 2. CHANNEL MODELS

It is a well known fact that the sound speed in the ocean is not constant. Actually, it changes, not only from point to point in the ocean, but at each point it shows yearly, monthly, and even daily fluctuations.

This fact determines a complex ray structure within the ocean, the rays being constantly refracted, due to the constantly varying refractive index. It is then possible for rays with different launching angles to reach the same distant point.

Frequently, the sound speed profile has a minimum at some inner point, between the surface and the bottom of the ocean; in this case, a duct is formed, the rays being trapped around the depth at which this minimum value occurs; the sound waves can then travel considerable distances with relatively small attenuation, when compared to the one predicted for a constant velocity medium.

In this section, several algorithms are presented, which predict the point to point ocean's transfer function for different models of the sound speed profile.

First, the simple case of a bilinear sound speed profile is analysed, in subsection 2.1. The results obtained are then extended, to include the special case of reflecting boundaries (subsection 2.2) and of mutilinear profiles (subsection 2.3).

The following simplifying assumptions are in force:

i) the ocean is a linear medium;

ii) there is horizontal homogeneity, namely, the sound speed is described by the velocity profile, which is a function of depth only;

iii) no random fluctuations are taken into account;

iv) the medium boundaries are plane and perfectly reflecting.

Under these hypotheses, and using either a bilinear or a multilinear model for the velocity profile, algorithms are given which determine the ray structure between two given points in the ocean.

A linear filter model for the ocean is developed as a parallel combination of several linear filters, each one representing an individual path. Along each path, it is assumed that the ocean's action can be modelled only by the introduction of a delay (corresponding to the propagation time) and the multiplication by a constant factor (describing the attenuation undergone in the propagation from source to receiver).

The impulse response of the ocean from point S to point R is:

$$h_{ch}(t,S,R)= \sum_{i=1}^{P(S,R)} b_i(S,R)\delta(t-d_i(S,R)) \tag{1}$$

where $P(S,R)$ is the number of paths joining S and R, and $d_i(S,R)$ and $b_i(S,R)$ are the delay and attenuation of path i. The impulse response is completely defined by the parameters $b_i(S,R)$ and $d_i(S,R)$, which in turn, can be shown to depend only on the launching angle of ray i (for a given (S,R) pair).

## 2.1 BILINEAR VELOCITY PROFILE

As a first order approximation to its characteristic variation in the deep ocean, the velocity of sound, $v(y)$ is considered to depend on depth, y, in the following way:

$$v(y) = \begin{cases} v_o + g_1(y- y_d), & y< y_d \\ v_o + g_2(y- y_d), & y\geq y_d \end{cases} \tag{2}$$

where $v_o$ is the minimum value of the sound speed, attained at depth $y_d$, and $g_1< 0$ and $g_2> 0$ are the gradient of $v(y)$ above and below $y_d$, respectively, see figure 1.

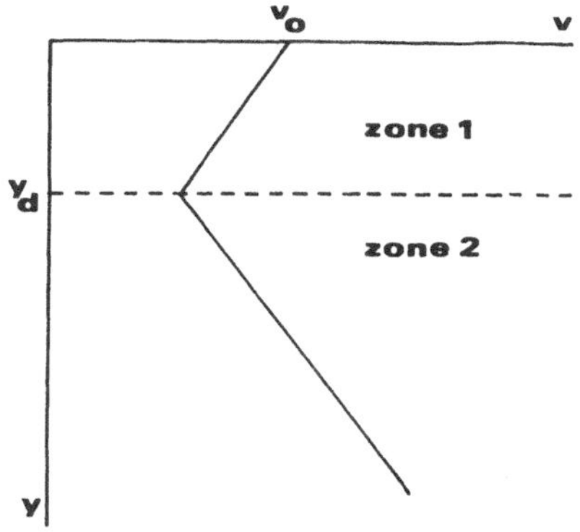


Figure 1 Linear Velocity Profile.

Figure 2 shows different types of rays that may occur propagating between source S and receiver R. The so called SOFAR rays (Ray I of figure 2) are purely refracted. All other rays have either bottom or surface reflections (e.g. Ray II), or multiple surface and bottom reflections (e.g. Ray III).

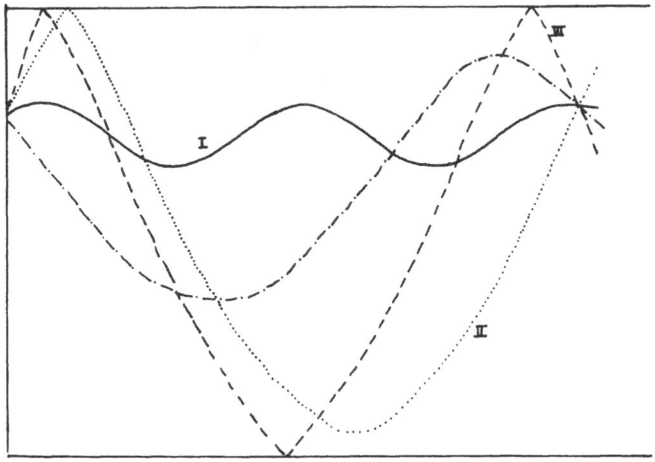


Figure 2 Ray Structure for a Nonconstant Velocity Profile.

It is well known that the ray paths in zones where the velocity gradient is constant are arcs of circle, whose parameters are determined entirely by the source depth $y_s$, and the launching angle, $\theta_s$. The continuity of the sound speed at $y=y_d$ determines the continuity of the tangent to the ray at that depth.

We first consider SOFAR rays. According to figure 3, the horizontal distance from S to R is,

$$x_r = x_{init} + K\, x_{hop} + \alpha\, x_i + x_{fin} \qquad (3)$$

where the meaning of the various parameters is defined below:

i) $x_{init}$ - horizontal distance from the source to the first crossing of the ray with the line $y=y_d$ (duct axis);

ii) $x_i$, i=1,2 - horizontal distance travelled in zone i (zone 1 corresponds to $y<y_d$ and zone 2 to $y \geq y_d$), between two consecutive crossings;

iii) $x_{hop} = x_1 + x_2$;

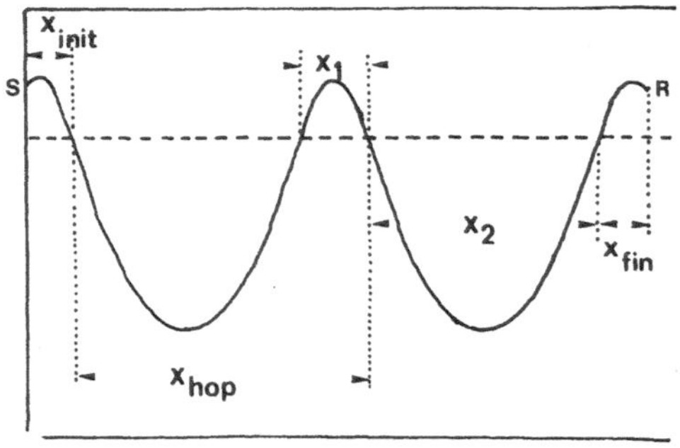


Figure 3 Decomposition of Rays.

iv) $x_{fin}$ - distance from last axis crossing to receiver;

v) K - number of complete hops;

vi) $\alpha$ - boolean variable; it takes the value 0 when source and receiver are in different zones, and the value 1 when they are in the same zone.

Eq.(3) is meaningful only when K >0 or K=0 and $\alpha$=1; when K=0 and $\alpha$=0, no crossing points exist, and no identification of the several terms in (3) is possible; this case (K=0, =0) will be considered separately.

I - <u>(K>0)</u> or <u>(K=0, $\alpha$=1)</u>

Expressing each distance as a function of the launching angle $\theta_s$, in (3), yields a 4th degree equation on the tangent of $\theta_s$. The four different solutions for $\theta_s$ reflect the fact that for each value of K there can be up to 4 distinct rays, corresponding to the different combinations of the following parameters:

i) $\beta = sgn(\theta_s)$

ii) $\delta = sgn(\theta_r)$

where $\theta_r$ is the receiving angle, and the signal function sgn(x) has the following definition:

$$sgn(x) = \begin{cases} 1, & x \geq 0 \\ -1, & x < 0 \end{cases}$$

The values of K that should be considered when solving (3) are:

$K \in \{K_{min}, \ldots, K_{max}\}$.

These limits are determined considering the rays which have the largest and smallest values of $x_{hop}$, for $K_{min}$ and $K_{max}$, respectively.

The ray with largest value of $x_{hop}$, is the ray which is tangent to the boundary where the sound speed is smaller, i.e., whose launching angle is determined by:

$v_{lim}$ := min (v(0), v(bottom));

$\theta_{min}$ := arcos($v_s/v_{lim}$);

The ray with smallest value of $x_{hop}$ is emitted with $\theta_s = \theta_{max}$, where

<u>if</u> ($v_s \leq v_r$) <u>then</u> $\theta_{max}$ := 0

<u>else</u> $\theta_{max}$ := arcos ($v_s/v_r$);

Algorithm (Determine K limits), given below, determines the limits for the parameter K, assuming the bilinear profile model.

Using geometric relations and Snell's law, the function $x_{hop}(\theta)$ is found to be [4]

$$x_{hop}(\theta) = x_1(\theta) + x_2(\theta)$$

$$= 2\, v_o(1/|g_1| + 1/|g_2|)[(v_s/v_o)^2(1+tg^2\theta) - 1]^{1/2}. \qquad (4)$$

The other distances in (3) are given by:

$$x_{init}(\theta) = (v_s/g_s)tg\theta + v_o[(v_s/v_o)^2(1+tg^2\theta) - 1]^{1/2}/|g_s| \qquad (5)$$

$$x_{fin}(\theta) = v_o[(v_s/v_o)^2(1+tg^2\theta) -1]^{1/2}/|g_r|$$

$$- v_r\delta[(v_s/v_r)^2(1+tg^2\theta)-1]^{1/2}/g_r \qquad (6)$$

where $g_s$ and $g_r$ are the velocity gradient at source depth and receiver depth, respectively.

<u>Algorithm</u> (Determine K limits)

<u>begin</u>

$v_{lim}$ := min (v(0), v(bottom));

$\theta_{min}$ := arcos($v_s/v_{lim}$);

<u>if</u> ($v_s \leq v_r$) <u>then</u> $\theta_{max}$:= 0 <u>else</u> $\theta_{max}$:= arcos($v_s/v_r$);

x_hop_max := $x_{hop}(\theta_{max})$;

x_hop_min := $x_{hop}(\theta_{min})$;

$$K_{min} := \left\lceil \frac{x_r}{\text{x\_hop\_min}} - \alpha \frac{|g_1| + |g_2|}{\min\limits_{i=1,2} |g_i|} - 2 \right\rceil;$$

$$K_{max} := \left\lfloor \frac{x_r}{\text{x\_hop\_max}} \right\rfloor;$$

<u>end</u>.

Symbol $\lceil x \rceil$ stands for the smallest integer greater than or equal to x, and $\lfloor x \rfloor$ for the largest integer smaller than or equal to x.

After lengthy algebraic manipulations, use of (4) to (6) in (3) yields the following equation on $tg\theta_s$:

$$a_o + a_1 tg\theta_s + a_2 tg^2\theta_s + a_3 tg^3\theta_s + a_4 tg^4\theta_s = 0 \qquad (7)$$

where:

$$a_o = B^2 - D \qquad (8.a)$$

$$a_1 = -4\ B\ x_r v_s/g_s \qquad (8.b)$$

$$a_2 = 2\ B\ C + 4\ x_r^2(v_s/g_s)^2 - E \qquad (8.c)$$

$$a_3 = -4\ C\ x_r v_s/g_s \qquad (8.d)$$

$$a_4 = C^2 - F^2 \qquad (8.e)$$

and,

$$B = x_r^2 \quad \backslash^2[(v_s/v_o)^2-1 + (v_r/g_r)^2((v_s/g_r)^2-1)] \qquad (9.a)$$

$$C = (v_s/g_s)^2 - (Av_s/v_o)^2 - (v_r/g_r)^2 \tag{9.b}$$

$$D = 4\ A^2 (v_r/g_r)^2 [((v_s/v_o)^2-1)((v_s/v_r)^2-1)] \tag{9.c}$$

$$E = 4\ A^2 (v_r/g_r)^2 [(v_s/g_r)^2 ((v_s/v_o)^2-1)+(v_s/v_o)^2 ((v_s/v_r)^2-1] \tag{9.d}$$

$$F = 4\ A^2 (v_r/g_r)^2 (v_s/v_r)^2 (v_s/v_o)^2 \tag{9.e}$$

$$A = v_o/|g_s| + 2\ v_o[K(1/|g_1| + 1/|g_2|) + \alpha/|g_i|] + v_o/g_r \tag{9.f}$$

II - <u>(K=0, α=0)</u>

In this case, the ray never crosses the duct axis. It is straightforward to show that:

$$tg\theta_s = \frac{g_s}{2v_s x_r} [x_r^2 + (y_r-y_d)^2 - (y_s-y_d)^2 - 2\ v_o(y_s-y_r)/g_s]. \tag{10}$$

The solutions of (7) and (10) should correspond to rays that are never reflected by the boundaries, that is, that satisfy:

$$|\theta_s| \leq \min\ (|\theta_{sur}|, |\theta_b|) \tag{11}$$

where

$$\cos\ \theta_{sur} = v_s/v(0)$$

$$\cos\ \theta_b\ = v_s/v(bottom).$$

Algorithm 1, given below, determines the set of source angles of SOFAR rays between two given points for a bilinear velocity profile.

Once the set of valid source angles is known, the delay from S to R along a ray with launching angle $\theta_s$ is:

$$d(\theta_s,K,S,R) = d_{init} + K\ d_{hop} + \alpha d_i + d_r. \tag{12}$$

Each term in (12) corresponds to the delay over each elementary path segment defined in (8), and is given by [4]

$$d_{init} = YB/|g_s|\ \ln \left|\frac{1 + [1 - \cos^2\theta_s]^{1/2}}{\cos\theta_s}\right| +$$

$$1/|g_s|\ \ln \left|\frac{1 + [1 - (v_o\cos\theta_s/v_s)^2]^{1/2}}{v_o\cos\theta_s/v_s}\right|$$

$$d_{hop} = 2(1/|g_1| + 1/|g_2|)\ \ln\left|\frac{1 + [1 -(v_o\cos\theta_s/v_s)^2]^{1/2}}{v_o\cos\theta_s/v_s}\right|$$

Algorithm 1 (Determine Source Angles)

```
begin
     input(S,R);     (* define Source and Receiver positions *)
     input(velocity profile);
     Determine K limits (K_min,K_max);
     P:=0;
     if (K_min= 0)
          then if (α = 0)  then begin
               solve Eq( [illegible] );
               if condition (11) then begin
                                        θ[P] := θ_s;
                                        P := P+1;
                                        end
               end
          else begin
               solve Eq(7);
               for each solution do
               if condition (11) then begin
                                        θ[P] := θ_s;
                                        P := P+1;
                                        end;
               end;
     for K := max(1,K_min) to K_max do begin
               solve Eq(7);
               for each solution do
               if condition (11) then begin
                                        θ[P] := θ_s;
                                        P := P+1;
                                        end;
               end;
     output(P,θ);
     end.
```

$$d_{fin} = (1-\mu\delta)/|g_r|\ \ln\left|\frac{1 + [1 - (v_r\cos\theta_s/v_s)^2]^{1/2}}{v_r\cos\theta_s/v_s}\right| +$$

$$1/|g_r|\ \ln\left|\frac{1 + [1 - (v_o\cos\theta_s/v_s)^2]^{1/2}}{v_o\cos\theta_s/v_s}\right|$$

where

$Y = \text{sign}(y_s - y_d)$

$\mu = \text{sign}(y_r - y_d)$.

The geometric spreading loss from S to R, for a ray emitted along $\theta_s$ is [4]:

$$I_o/I_r = |x_r\sin\theta_s/\cos\theta_s|\ \sin\theta_s.\{-v_r^2/(v_s g_r\cos^2\theta_r\sin\theta_r) +$$

$$v_s/(g_s\cos^2\theta_s\sin\theta_s) + Av_o/(v_s\cos^2\theta_d\sin\theta_d)\}$$

where $\theta_d$ is the angle between the tangent to the ray and the horizontal at depth $y_d$. The attenuation from S to R is:

$$b(\theta_s,S,R) = (I_o/I_r)^{-1/2} \tag{13}$$

Eqs.(12) and (13) define the filter's parameters, completing thus the modelling of the channel.

## 2.2 SHALLOW WATER CHANNEL

When the bottom depth is small, the predominant form of propagation is by successive reflections on the medium boundaries. In this section, a variation of the algorithm for the determination of the source angles previously presented is given, that solves the completely reflected rays case.

Eq.(3) of last section is still valid, the only difference being that each elementary path is no longer an arc of a circle, but the union of two, with a discontinuity point at one of the boundaries.

For $(K>0)$ and $(K=0,\alpha=1)$ the following equation on $tg\theta_s$ is obtained:

$$\begin{aligned} x_r = {} & A_1 tg\theta_s - A[(v_s/v_o)^2(1 + tg^2\theta_s) - 1]^{1/2} \\ & + B_1[(v_s/v(\text{bottom})^2(1 + tg^2\theta_s) - 1]^{1/2} \\ & + C_1[(v_s/v(0))^2(1 + tg^2\theta_s) - 1]^{1/2} \\ & + D_1[(v_s/v_r)^2(1 + tg^2\theta_s) - 1]^{1/2} \end{aligned} \tag{14}$$

where

$A_1 = v_o/g_s + y_s - y_d$

$B_1 = (v_o/g_1 + y_{bottom} - y_d)[-2K-½(Y+\beta)(1+Y) - ½(\mu-\delta)(1+\mu) - (1+Y)]$

$C_1 = (v_o/g_2 - y_d)[-2K-½(Y+\beta) - ½(\mu-\delta)(1-\mu) - (1-Y)]$

$D_1 = -(v_o/g_r + y_r - y_d)$.

The set of values of K to be considered when solving (14) is again determined using limiting arguments on the launching angle for these rays. Once it is assumed that the rays reflect in both boundaries, $\theta_s$ must satisfy:

$$|\theta_s| \geq \max(|\theta_{sur}|, |\theta_b|) \triangleq \theta_{lim} \qquad (15)$$

where the angles $\theta_{sur}$ and $\theta_b$ have been defined previously.

To the angle $\theta_{lim}$ corresponds the ray with smallest number of crossing points with the duct axis, yielding a lower limit on K:

$$K_{min} = \lceil x_r/x_{hop}(\theta_{lim}) - 1 - \alpha \rceil . \qquad (16)$$

The theorectical value for $K_{max}$ is infinity; however, a practical limit may be found, since for large values of K the attenuation is very large, making that path's contribution to the overall signal negligible.

The function $x_{hop}(\theta)$ appearing in (16), is given by

$$x_{hop}(\theta) = 2\, v_o(1/|g_1| + 1/|g_2|)[(v_s/v_o)^2(1+tg^2\theta) - 1]^{½}$$

$$- 2(v_o/|g_1| + y_d)[(v_s/v(0))^2(1 + tg^2\theta) - 1]^{½}$$

$$- 2(v_o/|g_2| + y_{bottom} - y_d)[(v_s/v_{bottom})^2 - 1]^{½}.$$

Eq.(14) is an 8th degree equation on $tg\theta_s$. It must be solved numerically for each value of K, to yield the source angles. The search interval for its resolution is restricted, for each K, from knowledge of the limit values that $x_{hop}$ can take.

The following relations hold:

$$x_{lim}^{K+2} \triangleq \frac{x_r}{K+2} < x_{hop}(\theta) < \frac{x_r}{K} \triangleq x_{lim}^{K} \qquad (17)$$

Solving both inequalities for θ, yields an upper and lower bound on its value.

Algorithm 2 (Determine Source Angles) calculates the set of valid launching angles of completely reflected rays for the bilinear velocity profile.

After the set of source angles is known, de[illegible]ination of the delay corresponding to each one parallels what has been done for completely refracted rays, in subsection 2.1; Eq(12) still holds, with the elementary delays 'eing now calculated integrating the delay along an infinitesimal path $d\tau$,

$$d\tau = \frac{dy}{v(y)[1 - v(y)^2/v_v]^{1/2}}$$

over each elementary path segment, defined between two consecutive inflection points. The quantity $v_v$ appearing in the previous equation is the vertex velovity:

$$v_v = v_s/\cos\theta_s.$$

Algorithm 2 (Determine Source Angles)

```
begin
     input(S,R);
     input (velocity profile);
     Determine K_min;              (* solve Eq(16) *)
     P := 0;
     for K := K_min to K_max do begin
          Determine Θ Interval;     (* solve (17) *)
          solve Eq(14);
          for each solution do
               if condition (15) then begin
                    θ[P] := θ_s;
                    P := P+1;
                    end;
          end;
     output(θ,P);
     end.
```

The determination of the attenuation is not, however, a simple generalization of the result of the last subsection, since now the losses due to boundary reflections must also be taken into account. In the hypothesis assumed (plane boundaries), and following [1], the losses due to boundary reflections can be modelled by:

$$\nu = \begin{cases} -1, & \text{for surface reflections} \\ e^{\beta_o}, & \text{for bottom reflections} \end{cases}$$

where $\beta_o$ is a parameter dependent on the bottom characteristics.

The total number of surface reflections $N_{sur}$, and bottom reflections $N_b$, are

$N_{sur} = 1/4(\beta\sigma+1)(Y-1) + K + 1/2\alpha(Y+1) + 1/4(1-\mu\sigma)(\mu-1)$

$N_b = 1/4(\beta\sigma+1)(Y+1) + K + 1/2\alpha(Y-1) + 1/4(1-\mu\sigma)(\mu+1)$

where $\beta$, $\alpha$, Y, $\sigma$ and $\mu$ have been defined previously.

The total attenuation due to reflections is therefore

$b_r = (-1)^{N_{sur}} . e^{N_b \beta_o}$

and the attenuation coefficient:

$b(S,R) = b_r . b_s$

where $b_s$ represents the geometric spreading loss.

## 2.3 MULTILINEAR PROFILE

In this section the velocity profile is modelled by a multilinear function of depth, satisfying the following conditions:

i) it is a continuous function of y;

ii) it exhibits an absolute minimum in an interior point $y_d$;

iii) its derivative dv/dy is a step function, nonincreasing for $y<y_d$, and nondecreasing for $y \geq y_d$.

Let

$$v(y) = v_{0,i} + g_i(y - y_i) \quad , i \in \{-M_2, \ldots, M_1\} \ , \ i \neq 0. \qquad (18)$$

In this expression, index <u>i</u> represents a zone of constant gradient; negative values of <u>i</u> being used for zones above the axis (understood as the line at which the velocity has its minimum value, $y=y_d$) and positive values for zones below. The other parameters in (18) are described below.

i) $M_1, M_2$ - number of intervals of linear variation of v(y), below and above the axis, respectively;

ii) $v_{0,i}$ lower value of v(y) in zone i;

iii) $g_i$ - velocity gradient in zone i;

iv) $y_i$ - upper limit of y in zone i.

It is considered, for the sake of simplicity, that $v_{0,i}=v_{0,-i}$. This assumption does not restrict the class of profiles considered, since a fictitious boundary may always be introduced, separating two zones with the same velocity gradient.

Going through the steps of subsection 2.1, the horizontal distance between S and R is decomposed as

$$x_r = x_{init} + K\, x_{hop} + \alpha\, x_i + x_{fin}. \tag{19}$$

Like before, the launching angle will be determined expressing each distance in (19) in terms of $tg\theta_s$. These expressions depend on the outermost zone reached by the ray. It is easily verified that rays with smaller values of $|\theta_s|$ are confined to zones closer to the axis than rays with larger values of $|\theta_s|$.

Let z>0 be the index of the outermost zone crossed by a ray. For a given (S,R) pair, z cannot be inferior to the index of the zones where the emitter and the receiver are ($i_s$ and $i_r$, respectively):

$$z \geq \max(|i_s|,|i_r|) \triangleq z_{min}. \tag{20}$$

In addition, since only completely refracted rays are considered,

$$z \leq \min(M_1,M_2) \triangleq z_{max}. \tag{21}$$

For each value of $z \in \{z_{min},\ldots,z_{max}\}$, expressions in terms of $\theta_s$ for $x_{init}$, $x_{hop}$, $x_i$ and $x_{fin}$ can be found. These expressions are quite involved, and consequently are not given here.

$$x_{init} = x_{s-} + \sum_{k=Y}^{Y(i_s-1)} x_k + (Y\beta+1)\left[x_{s+} + \sum_{k=Y(i_s+1)}^{y(z-1)} x_k + d_z\right]$$

with

$$x_{s-} = (|y_{i_s}| + |y_c|)tg\theta_i - (|y_c| + |y_s|)tg\theta_s$$

$$x_k = (|y_k| + |y_{c,k}|)tg\theta_k - (|y_{k+1}| + |y_{c,k}|)tg\theta_{k+1}$$

$$\theta_k = \arccos\left[(v_{o,k}/v_s)\cos\theta_s\right]$$

$$y_{c,k} = v_{0,k+1}/g_k$$

$$x_{s+}=\begin{cases} 2(v_{0,i_s}/|g_s| + |y_s|)tg\theta_s , & i_s=z \\ \beta(v_{o,i_s}/|g_s| + |y_s|)tg\theta_s - (v_{o,i_s}/|g_s|+|y_s|)tg\theta_{i_s+1}, & i_s\neq z \end{cases}$$

$$d_z = \begin{cases} 0 , & i_s= z \\ 2(v_{o,z}/|g_z| + |y_z|)tg\theta_z , & i_s\neq z. \end{cases}$$

The distance between two consecutive crossing points with the axis is

$$x_i = d_{-Yz} + 2 \sum_{k=-Y}^{-Y(z-1)} x_k ,$$

while

$$x_{hop} = d_z + d_{-z} + 2 \sum_{k=-1}^{-(z-1)} x_k + 2 \sum_{k=1}^{z-1} x_k .$$

Finally,

$$x_{fin} = x_{i_r-} + \sum_{k=\mu}^{\mu(i_r-1)} x_k + (1-\mu\acute{o})[x_{i_r+} + \sum_{k=\mu(i_r+1)}^{\mu(z-1)} x_k + d_z],$$

where:

$$x_{i_r-} = (|y_{i_r}| + |y_{c,i_r}|)tg\theta_{i_r} - (|y_r| + |y_{c,i_r}|)tg\theta_r$$

$$x_{i_r+}=\begin{cases} 2(|y_r| + |y_{c,i_r}|)tg\theta_r & i_r=z \\ (|y_r| + |y_{c,i_r}|)tg\theta_r - (|y_r| \quad |y_{c,i_r}|)tg\theta_{i_r+1}, & ir=z \end{cases}$$

The procedure (Determine K Interval), stated below, finds, for each

$z \in \{z_{min},...,z_{max}\}$

the interval of variation of K. In this procedure, $\theta_{lim,j}$ is the launching angle of the ray tangent to the line $y=y_j$:

$$\cos \theta_{lim,j} = v_s/v_{0,j}$$

for $j \in \{-M_2,...,M_1\}$.

Algorithm 3, presented after procedure (Determine K Interval), identifies the set of source angles of SOFAR rays (assuming a velocity profile described by a multilinear function). Determination of the attenuation and delay for each ray is identical to what has been done for the two models previously studied.

<u>Algorithm</u> (Determine K Interval)

```
begin
     if (|i_s| > |i_r|) then
          if (z = |i_s|) then begin
               K_min := x_r/x_hop(θ_lim,i_s) ;
               K_max := x_r/x_hop(0) - 2 ;
               end
          else begin
               K_min := ⌈x_r/x_hop(θ_lim,z+1) -2⌉;
               K_max := ⌊x_r/x_hop(θ_lim,z)⌋ ;
               end;
     end.
```

The algorithm that computes the source angles can finally be stated:

<u>Algorithm 3</u> (Determine Source Angles)

```
begin
     input (S,R);
     input (velocity profile);
     Determine z interval            (*Eqs(20) and (20) *)
     for z := z_min to z_max do begin
          Determine K interval;
          for K := K_min to K_max do begin
               solve (19)
               for each solution do
                    if (|θ_s|<θ_lim,z+1)and (|θ_s|>θ_lim,z) then
                         begin
                         θ[P] := θ_s;
                         P := P+1;
                         end;
               end;
          end;
          output(P,θ);
     end.
```

## 2.4 EXAMPLES

To illustrate the type of impulse responses that can be expected, algorithm 1 is applied, for different relative positions of the emitter and receiver.

The channel's parameters used were:

$g_1 = -.0306\ s^{-1}$ $\quad g_2 = .01601\ s^{-1}$

$y_d = 914$ m $\quad$ bottom = 3714 m $\quad v_0 = 1480$ m/s

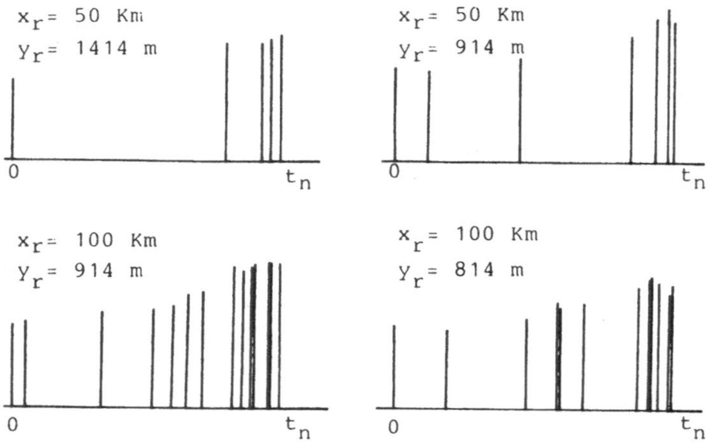


Figure 4 Impulse Responses (Deep Ocean).

In figure 4, the normalized impulse response for 4 different (S,R) pairs is illustrated. Note that in this figure, the time and attenuation scales have been normalized:

$$t_n = (t - \tau_{min}) / d\tau_{max}$$

where

$$d\tau_{max} = \max_{i \neq j} |d_i - d_j|,$$

$$\tau_{min} = \min_i d_i,$$

and

$$a_n = a / a_{max}$$

where

$$a_{max} = \max_i a_i .$$

General conclusions that can be withdrawn from this figure are

i) as the horizontal distance from S to R increases, the number of SOFAR rays increases,

ii) in general, the replicas that suffer greatest attenuation (smaller values of $a_i$) are the least delayed ones. It can be shown that they correspond to propagation paths that get closer to the boundaries, i.e., that have the larger values of $|\theta_s|$,

iii) analysis of the eigenray plots as in figure 4 may help in actual situations to decide on how to cluster them, with the ensueing receiver simplifications.

## 3. MAXIMUM LIKELIHOOD ESTIMATOR OF SOURCE LOCATION IN THE PRESENCE OF MULTIPATH

The source location problem can be stated as a particular instance of the general problem of parameter estimation in stochastic signals:

Problem

Given

$$r(t)=s_r(t:a) + w(t), \quad t \in [T_i,T_f],$$

where $s_r(t:a)$ is a function of the unknown parameter vector a :

$$s_r(t:a) = \sum_{k=1}^{P} b_k(a)\, s(t-d_k(a)) \tag{23}$$

and:

- $s(t)$ is a sample function of the source signal process;
- $w(t)$ is a sample function of white measurement noise, of known spectral density $N_0/2$, independent of $s(t)$;
- $b_k(a)$ and $d_k(a)$ , $k=1,\ldots,P$ are deterministic functions of the parameter vector a.

Determine the maximum likelihood estimate (MLE) of a, i.e., the value $\hat{a}_{ML}$ that maximizes the conditional probability density function:

$$p(r(t),\ t\in[T_i,T_f] \mid a).$$

The expression for $s_r(t:a)$ assumes the multipath structure. According to the velocity profile considered, the delays $d_k(a)$ and attenuations $b_k(a)$, are given by one of the expressions on the previous section. The parameter vector a describes the source position in the chosen coordinate system.

It is further assumed that the source signal s(t) is a sample function of a Gaussian process, with zero mean and known covariance function. In this case, it is shown that the MLE maximizes the log-likelihood function (LLF), given by [5]:

$$LLF(a)=1/N_0\int_{T_i}^{T_f} r(t)\ s_r^{\star}(t:a)dt - ½\int_0^{2/N_0} dw \int_{T_i}^{T_f} h(t,t:a|w)\ dt \qquad (24)$$

where:

- $s_r^{\star}(t:a)$ is the unrealizable minimum mean square error estimate (MMSEE) of $s_r(t,a)$;
- $h(t,v:a|w)$ is the impulse response of the optimum unrealizable filter that yields $s_r^{\star}(t:a)$, assuming a is known, and for measurement noise of spectral height w.

An alternative expression for the LLF is obtained, which makes use of the realizable (causal) filtering estimate, $\hat{s}_r(t|t:a)$:

$\hat{s}_r(t|t:a) = E[r(t)|r(s),T_i \leq s < t,a]$

The function to be maximized is [5]

$$LLF(a) = 1/N_o\int_{T_i}^{t} [2r(u)\hat{s}_r(u|u:a) - \hat{s}_r^2(u|u:a)]du$$

$$- 1/N_o\int_{T_i}^{t} \xi_{P_s}(u)\ du \qquad (25)$$

where $\xi_{P_s}(u)$ is the minimum mean square error of the causal estimate.

The parameter estimation problem involves, in either case, the resolution of a related waveform estimation problem; i.e, the LLF is made to depend either on the solution of an optimal smoothing problem (determine $s_r^{\star}(t:a)$) or on the solution of an optimal filtering problem (determine $\hat{s}_r(t|t:a)$). On the sequel, attention is given to each of these signal estimation problems, for the special case of multipath propagation.

3.1 SMOOTHING PROBLEM

In multipath ambient, where the information bearing component of the received signal is the superposition of a discrete number P of distorted replicas of the emitted signal, the optimum filter structure ressembles that assumed for the channel. Namely, as it will be shown, the optimum filter is naturally decomposed in P subfilters, each one dedicated to the estimation of an individual component of $s_r(t:a)$. Furthermore, each one of these filters is still decomposed in P blocks, reflecting the structure assumed for the incoming signal.

The equation defining the optimum unrealizable filter h(t,u:a) is [5]:

$$\int_{T_i}^{T_f} h(t,u:a)K_r(u,v:a)\,du = K_{s_r}(t,v:a) \tag{26}$$

where

$$K_r(u,v:a) = E[r(u)r(v)|a]$$

is the covariance of the received signal; and

$$K_{s_r}(u,v:a) = E[s_r(u)s_r(v)|a]$$

is the covariance function of the information bearing component $s_r(t)$.

In multipath ambient,

$$s_r(t:a) = \sum_{k=1}^{P} h_{ch_k}(t:a)*s(t) \tag{27}$$

each $h_{ch_k}(t:a)$ representing the impulse response of the channel over path k, for an emitter described by the vector a.

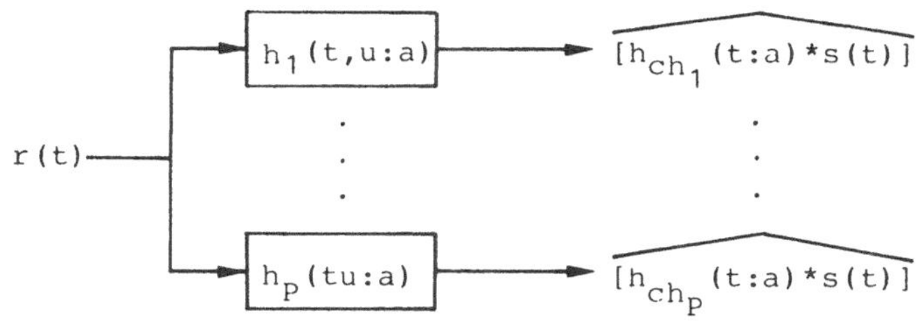


Figure 5 Decomposition of Optimal Smoother.

Using this expression on the definition of $K_{s_r}(u,v:a)$, manipulations of the integral equation (27) lead to the following decomposition of the optimum filter:

$$h(t,u:a) = \sum_{k=1}^{P} h_k(t,u:a) \qquad (28)$$

where $h_k(t,u:a)$ is the optimum unrealizable filter to estimate the replica received over path k, assuming a is known. The struture obtained is represented in figure 5.

According to the model developed in the previous section, all paths' impulse responses are identical, differing only on the delay and attenuation introduced, and consequently, all the filters in figure 5 satisfy identical equations.

Further decomposition of each $h_k(t,u:a)$ in P components is achieved, the ith component of filter $h_j(t,u:a)$ being the optimum unrealizable filter for the estimation of

$$d_{ij}(t:a) = h_{ch_j}(t:a) * h_{ch}^{-1}(t:a) * h_{ch_i}(t:a) * s(t) \qquad (29)$$

where $h_{ch}^{-1}(t:a)$ stands for the impulse response of the inverse filter of $h_{ch}(t:a)$.

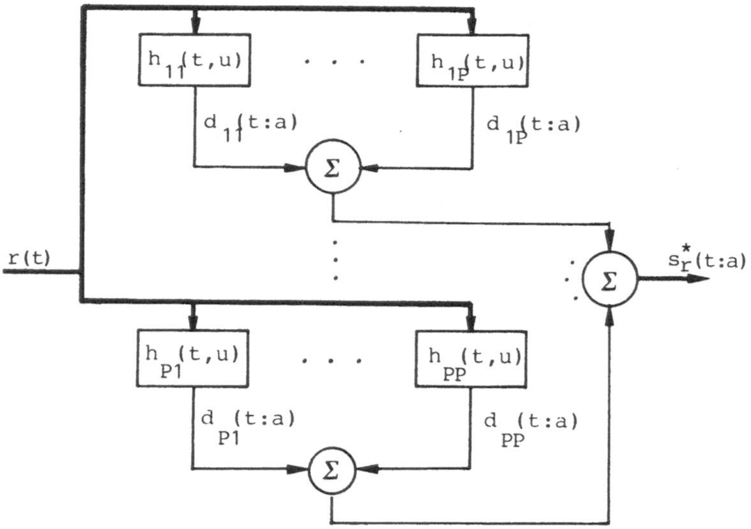


Figure 6 Optimal Smoother Structure.

Obtaining Eq(29) involves the utilisation of the inverse kernel of $K_r(t,u:a)$ on the integral equation defining $h_j(t,u:a)$, and its decomposition on a form similar to that of $K_s(t,u:a)$; the details can be found in [4]. The final structure is illustrated in figure 6. It ressembles that proposed in [6] for the multiple source problem. However, in that paper, a fundamental assumption is that no source signal is an attenuated and delayed replica of any other source, and consequently is not aplicable for the model of multipath used here.

For the particular instance of stationary signals, and long observation interval, a simplified structure is obtained, similar to the classical beam-former [4]. However, the classical beam-former is derived for the estimates of the intersensors' delays, and here the delays are due, not to a spatial sampling of the observed field, but to the temporal diversity of the channel.

## 3.2 FILTERING PROBLEM

The structure of the optimum causal filter for the multipath problem is, in general, quite complicated. Actually, due to the structure of the incoming signal, the filtering problem results in a mixture of smoothing, filtering and prediction problems.

In the filtering problem, we dealt with the calculation of

$$\hat{s}_r(t|t:a) = E[r(t)|r(s),\ T_i \leq s < t,\ a].$$

According to (23),

$$\hat{s}_r(t|t:a) = \sum_{i=1}^{P} b_i(a)\hat{s}(t-d_i(a)|t:a).$$

At first sight, we are faced with P smoothing problems: determine $\hat{s}(t-d_i(a)|t:a)$, $i=1,\ldots,P$.

Let us write the received signal as:

$$r(t) = s_i(t) + \sum_{k<i} b_k(a)s(t-d_k(a)) + \sum_{k>i} b_k(a)s(t-d_k(a)) + w(t)$$

and consider $s_i(t) = b_i(a)s(t-d_i(a))$ the information bearing signal, and the other replicas as "noise" perfectly correlated with the signal. Let the desired signal be

$$d(t) = s_i(t-D).$$

The following three cases may occur:

i) $D = 0$ : the estimation problem is formulated as a filtering problem;

ii) $D = d_j(a) - d_i(a) > 0$ : this formulation corresponds to a smoothing problem;

iii) $D = d_j(a) - d_i(a) < 0$ : we are faced with a prediction problem.

This mixture of the three types of waveform estimation problems makes it very difficult to obtain results for the general case. In a following subsection, the use of a state model for the source signal will be shown to allow for the determination of the optimum filter structure.

### 3.3 OPTIMAL SMOOTHER - Source signal in the subspace spanned by a known set of functions.

In this and the next subsection, we restrict the class of source signals being considered, and present the resulting filter structure.

On the sequel, the optimal smoother structure is developed, assuming the following additional hypothesis:

$H_1$: The source signal s(t) belongs to the subspace spanned by the linearly independent functions $f_i(t)$, $i=1,\dots,M$:

$$s(t) = \sum_{i=1}^{M} s_i f_i(t) \tag{30}$$

where $s_i$, $i=1,\dots,M$ are random coefficients, of known covariance matrix:

$$\Lambda = E[ss'].$$

It is well known that [5] the optimum unrealizable filter can be expressed in terms of the eigenvalues $\lambda_i(a)$ and eigenfunctions $\Phi_i(t{:}a)$ of the received signal covariance function:

$$h(t,u{:}a) = \sum_{i=1}^{M} \lambda_i(a)[\lambda_i(a)+N_o/2]^{-1}\Phi_i(t{:}a)\ \Phi_i(u{:}a). \tag{31}$$

Under $H_1$, $K_{s_r}(t,u{:}a)$ can be expressed in terms of the functions $\{f_i(t)\}$ and of the matrix $\Lambda$:

$$K_{s_r}(t,u:a) = \sum_{i,j=1}^{P} b_i(a)b_j(a)\sum_{k,m=1}^{M}\Lambda_{km}f_k(t-d_i(a))f_m(t-d_j(a))$$

or, more compactly:

$$K_{s_r}(t,u:a) = g'(t:a)\Lambda g(u:a) \tag{32}$$

where g(t:a) is an (Mx1) vector, whose ith component is

$$g_i(t:a) = \sum_{k=1}^{P} b_k(a)f_i(t-d_k(a)). \tag{33}$$

This is the vector of transformed basis functions, and it contains all the information that is needed to know the source location, i.e., the parameter vector a.

Even if the functions $\{f_i(t)\}$ are linearly independent, we cannot guarantee that the components of g(t:a) are. Let $\{\Psi_i(t:a)\}$, $i=1,\ldots,N\leq M$, be a set of orthonormal functions, obtained from $\{g_i(t:a)\}$ by the Gram-Schmidt method, and P(a) the matrix that transforms $\{\Psi_i(t:a)\}$ in $\{g_i(t:a)\}$:

$$g(t:a) = P(a)\Psi(t:a).$$

In terms of $\Psi(t:a)$,

$$K_{s_r}(t,u:a) = \Psi'(t:a)Q(a)\Psi(u:a) \tag{34}$$

where it has been defined

$$Q(a) = P'(a)\Lambda P(a).$$

Let $\rho_i(a)$ be an eigenvalue of Q(a), with normalised eigenvector $x^i(a)$. It can be shown that $\rho_i(a)$ is an eigenvalue of $K_{s_r}(t,u:a)$, i.e., $\lambda_i(a) = \rho_i(a)$, with corresponding eigenfunction:

$$\Phi_i(t:a) = x^i(a)'\Psi(t,a).$$

The eigenvectors of Q(a), $x^i(a)$, are the coefficients of the eigenfunctions $\{\Phi_i(t:a)\}$ on the orthonormal basis $\{\Psi_i(t:a)\}$.

From the previous relations,

$$h(t,u:a) = g^{(N)}(t:a)'H^{(N)}(a)g^{(N)}(u:a) \tag{35}$$

where $g^{(N)}(t:a)$ is a (Nx1) vector, formed by N linearly independent components of g(t:a), and:

$$H^{(N)}(a) = P^{(N)}(a)^{+}[I + (N_o/2)Q^{-1}(a)]P^{(N)}(a)^{-1}$$

where $A^{+}$ denotes the transpose of the inverse matrix of A.

When M=N, i.e., when all the functions $\{g_i(t:a)\}$ are linearly independent,

$$g^{(N)}(t:a) = g(t:a)$$

and,

$$H^{(N)}(a) = [T(a) + N_0/2\ \Lambda^{-1}]^{-1}$$

where T(a) is the Gram matrix of the functions $\{g_i(t:a)\}$:

$$T(a) = \langle g(t:a), g'(t:a) \rangle.$$

From now on, it will be assumed that M=N. The case M>N can be similarly treated, if $K_{s_r}(t,u:a)$ is expressed in a subset $(g^{(N)}(t:a))$ of the functions $\{g_i(t:a)\}$. The superscript N will also be dropped.

For N=M, it is easily shown that H(a) is the matrix that maps the vector of sufficient statistics:

$$u(a) = \langle r(t), g(t:a) \rangle$$

on the MAP (Maximum A Posteriori) estimate of the coefficients $s_i$:

$$\hat{s}_{MAP}(a) = H(a)\ u(a).$$

Finally, the estimate of $s_r(t:a)$ is:

$$\begin{aligned} s_r^*(t:a) &= \int_{T_i}^{T_f} h(t,u:a)\ r(u)\ du \\ &= g(t:a)'H(a)u(a) \\ &= g(t:a)'\hat{s}_{MAP}(a). \end{aligned}$$

The estimation process is: firstly, the inner product of r(t) and the received signal's basis functions is made, yielding the vector of sufficient statistics u(a); secondly, the matrix H(a) is applied to u(a), producing $\hat{s}_{MAP}(a)$, which, in turn, is combined with the known basis g(t:a), generating the MMSEE of $s_r(t:a)$. This process is illustrated in figure 7.

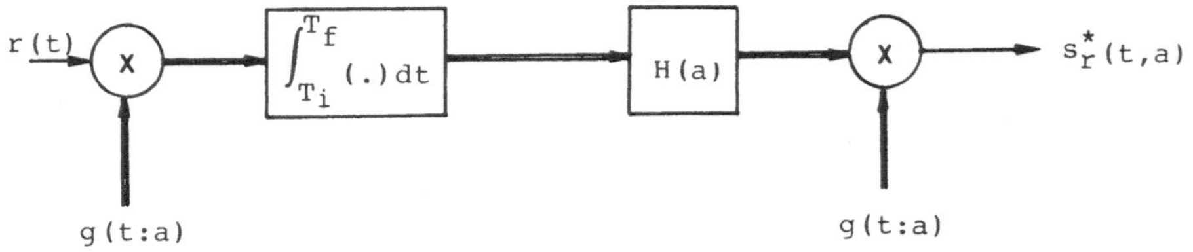


Figure 7 Optimal Smoother for the Degenerated Kernel Case.

From (35) it follows that this filter admits the decomposition in a PxP matrix of subfilters presented in section 3.1.

In order to estimate the parameter vector a, this estimation procedure is repeated for values of a in the admissible set, yielding the function LLF(a). The value of a for which LLF(a) is maximum is the MLE of a, $\hat{a}_{ML}$.

### 3.4 OPTIMAL FILTER - Source signal as the output of a finite dimensional linear system driven by white noise.

As it has been pointed out, the optimal filtering problem degenerates in a set of P smoothing problems. In this section the smoothing problem is assessed, under the following hypothesis on the source signal:

$H_2$ : The source signal, s(t), admits the description:

$$\dot{x}(t) = A(t)x(t) + u(t)$$
$$x(t_o) = x_o$$
$$s(t) = C(t)x(t)$$

where x(t) is an n-dimensional vector, A(t) is an nxn matrix, u(t) is a sample function of white Gaussian noise with zero mean and spectral height K, and $x_o$ is a Gaussian random vector, with zero mean and covariance matrix $P_o$. The source signal s(t) is scalar, and consequently C(t) is an (1xn) vector.

The received signal, under $H_2$, is:

$$r(t) = \sum_{k=1}^{P} q_k(t:a)\ x(t-d_k(a)) + w(t) \tag{36}$$

where it has been defined the (1xn) vector

$$q_k(t:a) = b_k(a)C(t-d_k(a)).$$

The problem of determining the MMSEE of x(t-d) from observations of r(t) of the form (36) has been solved in [3], for the general case of systems with delays in the state and observation equations. It is shown that $\hat{x}(t-d|t:a)$ satisfies the partial differential equation:

$$\frac{\partial \hat{x}(t-d|t:a)}{\partial t} + \frac{\partial \hat{x}(t-d|t:a)}{\partial d} = h_r(t,t:d:a).$$
$$[r(t) - \sum_{i=1}^{P} q_i(t:a)\hat{x}(t-d_i(a)|t:a)]$$

whose initial condition is the filtering equation:

$$\frac{\partial \hat{x}(t|t:a)}{\partial t} = A(t)\,\hat{x}(t|t:a) +$$

$$h_r(t,t:0:a)[r(t) - \sum_{i=1}^{P} q_i(t:a)\hat{x}(t-d_i(a)|t:a)].$$

Note that this is very similar to the Kalman-Bucy filter equation, the innovations being now computed using the smoothing estimates.

The gain of the differential equation, $h_r(t,t:d:a)$ is related to $P(t,d_1,d_2:a)$, the estimation error covariance, by:

$$h_r(t,t:d:a) = 2/N_0 \sum_{j=1}^{P} P(t,d,d_j(a))q_j(t:a)$$

which, in turn, is described by the nonlinear partial differential equation:

$$\frac{\partial P(t,d_1,d_2:a)}{\partial t} + \frac{\partial P(t,d_1,d_2:a)}{\partial d_1} + \frac{\partial P(t,d_1,d_2:a)}{\partial d_2} =$$

$$= 2/N_0 \sum_{i,j=1}^{P} P(t,d_1,d_i(a):a)q_i(t:a)'q_j(t:a)P(t,d_j(a),d_2:a) \qquad d_1,d_2 \geq 0$$

with the following boundary conditions:

$$\frac{\partial P(t,d_1,0:a)}{\partial t} + \frac{\partial P(t,d_1,0:a)}{\partial d_1} = P(t,d_1,0:a)\ A(t)'$$

$$- 2/N_0 \sum_{i,j=1}^{P} P(t,d_1,d_i(a):a)q_i(t:a)'q_j(t:a)P(t,d_j(a),0:a), d_1 \geq 0$$

$$\frac{\partial P(t,0,d_2:a)}{\partial t} + \frac{\partial P(t,0,d_2:a)}{\partial d_2} = A(t)\ P(t,0,d_2:a)$$

$$- 2/N_0 \sum_{i,j=1}^{P} P(t,0,d_i(a):a)q_i(t:a)'q_j(t:a)P(t,d_j(a),d_2:a), d_2 \geq 0$$

$$\frac{\partial P(t,0,0:a)}{\partial t} = A(t)\ P(t,0,:a) + P(t,0,0:a)A(t)' + R$$

$$+ 2/N_0 \sum_{i,j=1}^{P} P(t,0,d_i(a):a)q_i(t:a)'q_j(t:a)P(t,d_j(a),0:a)$$

The covariance equation does not depend on the data, and can be integrated prior to the estimation process, to yield the gain $h_r(t,t:d:a)$, for $d$ in the interval $[\min_i d_i(a), \max_j d_i(a)]$. For each particular value of a, there is an optimal gain, which, used in the integration of (37) will produce the MMSE estimates $\hat{x}(t-d|t:a)$.

The received signals' estimates are:

$$\hat{s}(t-d_i(a)|t:a) = C(t-d_i(a))\hat{x}(t-d_i(a)|t:a).$$

Finally, the information signal's estimate is reconstructed:

$$\hat{s}_r(t|t:a) = \sum_{i=1}^{P} b_i(a)\hat{s}(t-d_i(a)|t:a).$$

Even though this estimate can be recursively obtained, the source location vector estimate must resort to the maximization of LLF(a), given by (25), over the whole admissible set, and cannot be made recursive.

The overall estimation procedure is illustrated in figure 8.

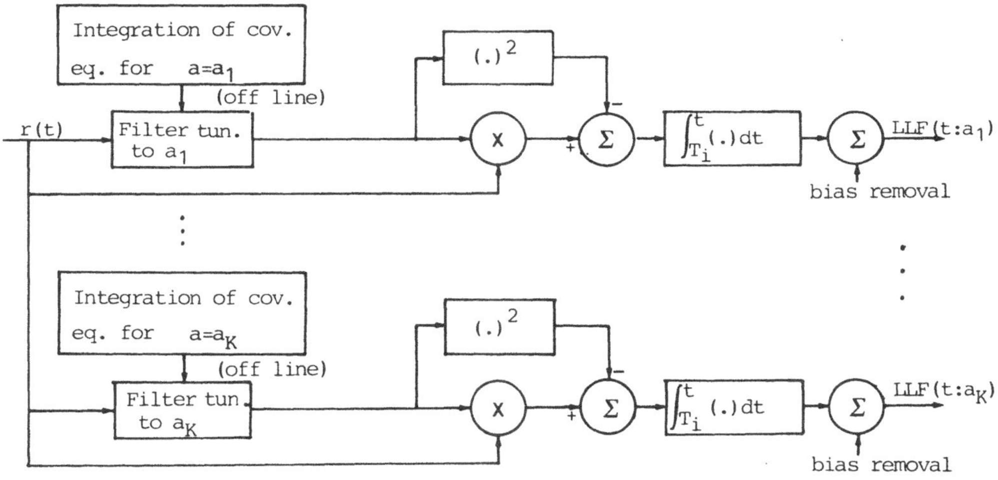


Figure 8 Block Diagram of the Estimator of Source Position.

This topic is not further pursued here. The application of the above filter to delay estimation problems has been done in [2].

## 4. CONCLUSIONS

In a variety of signal processing applications, of which the underwater acoustic is paradigmatic, the propagation characteristics of the medium enrich the received signal with a multipath structure. By assuming a multilinear velocity profile, the paper has presented in section 2 an algebraic algorithm that determines the number P of rays connecting the source to the receiver and the associated delays and attenuations. This algorithm uses the velocity profile , the source and receiver positions, and the medium configuration. As the source/receiver (horizontal) distance increases, the ray path structure becomes increasingly complex and a (normal) mode analysis is more appropriate. Nevertheless, because of its simplicity, the algorithm described is still useful, providing a first cut determination of the installed rays. The second question dealt with was concerned with the impact of the multipath signal on the receiver. The decompositions studied help comprehend the receiver's behaviour. This understanding is deepened by conveniently modelling the source radiated signal. Two different classes of signals, one where the signal is given as a linear combination of a finite number of basis functions, and the other when it is the output of linear finite dimensional filters, were finally considered.

## TABLE OF SYMBOLS

| | |
|---|---|
| $S$ - | source position. |
| $R$ - | receiver position. |
| $g$ - | velocity gradient. |
| $y$ - | depth (measured from surface, downwards). |
| $x_r$- | horizontal distance from source to receiver. |
| $y_s$- | source depth. |
| $y_r$- | receiver depth. |
| $v$ - | velocity of sound waves. |
| $g_s$- | velocity gradient at source depth. |
| $g_r$- | velocity gradient at receiver depth. |
| $v_s$- | velocity at source depth. |
| $v_r$- | velocity at receiver depth. |
| $v_o$- | minimum value of sound speed. |
| $\theta_s$- | launching angle. |
| $\theta_r$- | receiving angle. |
| $\theta_{sur}$- | launching angle of ray tangent to the surface. |
| $\theta_b$- | launching angle of ray tangent to the bottom. |
| $y_d$- | depth of minimum velocity. |
| $\alpha$ - | indicates whether source and receiver are in the same zone ($\alpha$=1) or in different zones ($\alpha$=0). |

$\beta$ - indicates whether the launching angle is positive ($\beta$=1) or negative ($\beta$=-1).

$\acute{o}$ - indicates whether the receiving angle is positive ($\acute{o}$=1) or negative ($\acute{o}$=-1).

Y - indicates whether the source is located above (Y=-1) or below (Y=1) the duct axis.

$\mu$ - indicates whether the receiver is below ($\mu$=1) or above ($\mu$=-1) the duct axis.

P - number of paths (rays) joining source and receiver.

$d_i$- delay along path i.

$b_i$- attenuation along path i.

a - parameter vector describing the source position.

r(t) - received signal.

$s_r(t)$- information bearing component of the received signal.

$\hat{s}_r(t)$- optimal filtering estimate of $s_r(t)$.

$s_r^*(t)$- optimal smoothing estimate of $s_r(t)$.

$h_{ch}(t)$ impulse response of the ocean.

$h_{ch_i}(t)$ impulse response of the ocean along path i.

$\{f_i(t)\}$ set of known basis functions of the source signal.

s - vector of expansion coefficients of source signal.

$\Lambda$ - covariance matrix of the vector of coefficients s.

$\{g_i(t)\}$ set of transformed basis functions $g_i(t)=h_{ch}(t)*f_i(t)$.

$\{\Psi_i(t)\}$ set of orthonormal functions obtained from $\{g_i(t)\}$.

Q covariance matrix of the expansion coefficients of $s_r(t)$ on the orthonormal basis $\{\Psi_i(t)\}$.

$K_{s_r}(t,u)$ covariance function of $s_r(t)$.

$\lambda_i$- eigenvalue of $K_{s_r}(t,u)$.

$\Phi_i(t)$ eigenfunction of $K_{s_r}(t,u)$.

ACKNOWLEDGEMENTS

This research was partially supported by INIC (Portugal).

## 5. REFERENCES

1. Gerlach, Acoustic transfer function of the ocean for a motional source, I.E.E.E Transactions on Acoustic, Speech and Signal Processing, Vol.ASSP-26, No.6, Dec 1978.

2. I.M.Lourtie and J.M.Moura, Time delay determination: maximum likelihood and Kalman-Bucy type structures, Proceedings of Int. Conf. Acoust. Speech and Signal Proc., 1985.

3. Kwakernaak, Optimal filtering in linear systems with time delays, IEEE Transactions on Automatic Control, Vol AC-12, No.2, April 1967.

4. M.J.Rendas, Estruturas óptimas para localização passiva em meios com dispersão temporal, Thesis, Depart. Eng. Electr. Comput., I.S.T, 1984.

5. H.L.Van Trees, Detection Modulation and Estimation, vol.III, New York, Willey, 1968.

6. M. Wax, T. Kailath, Optimum location of multiple source by passive arrays, IEEE Transactions on Acoustics Speech and Signal Processing, Vol.ASSP-31, No.5, Oct. 1983.

CHAPTER 4

# LEVEL CROSSING REPRESENTATIONS, POISSON ASYMPTOTICS AND APPLICATIONS TO PASSIVE ARRAYS

A.O. HERO S.C. SCHWARTZ

## I. INTRODUCTION

In this chapter we present representation theorems on the distribution of level crossing statistics for general non-stationary random processes and then discuss applications to passive arrays in underwater acoustics. Along with some regularity conditions, a consequence of the first representation theorem is an asymptotic result relating the distribution of the level crossings to an inhomogeneous Poisson law. This asymptotic result is then used to model the occurrence of large error in time difference of arrival estimates across a passive sensor array.

The distribution function of the number of crossings of a level by a random process in a given interval is an important, but difficult to obtain, function which has received considerable attention in recent years [1]. Explicit results are known only for a handful of specific random processes, e.g. the Markov class. Here a general representation for the probability of getting one or more level crossings is presented for a wide class of random processes, which allows the deviation of this probability from the probability of getting one or more points from an inhomogeneous Poisson process to be characterized. This representation is then exploited to obtain a generalization of asymptotic results of Leadbetter [12] for stationary random processes to the non-stationary case. The most familiar non-precise statement of Leadbetters' result is as follows. Define a suitable sequence of increasingly high levels $\{l_m\}$. Given a stationary, almost surely continuous random process $X$ which satisfies some regularity conditions, the counting process $N^m$ associated with the crossings by $X$ of the increasingly high level $l_m$, behaves increasingly like a (homogeneous) Poisson process. An equivalent interpretation: if $\mu_m$ and $\sigma_m^2$ are the mean and variance of $X_m$, which is one in a sequence of processes $\{X_m\}$, the Poisson character of its zero upcrossings, $N^m$, is assured as $\mu_m / \sigma_m \rightarrow \pm\infty$.

Here we establish an analogous, but inhomogeneous, Poisson character for non-stationary processes satisfying some additional conditions to those in [12]. Specifically, we assume certain asymptotic conditions on the trajectories of the process such as mixing. It is demonstrated that if the upcrossings of zero are made to become progressively rarer events, in a sense to be made clear later, then a normalized version of the number of upcrossings as a function of time converges in distribution to a Poisson process.

Results of the above type are of interest in connection with maximum likelihood parameter estimation when large errors may be significant. As is well known, the Cramer-Rao-Lower-Bound only characterizes local error; that is, when the estimate is in the immediate vicinity of the true parameter [4]. When the trajectory of the so called "likelihood function" is predisposed to display large multiple local maxima over the parameter space, or peak ambiguities, an additional large error measure is needed. One possible choice is the probability that a section of the trajectory exceeds a threshold, specified by the height of the trajectory at the true parameter. This probability can then be expressed within the framework of level crossing probabilities.

For time delay estimation in a two sensor array, the maximum likelihood estimate is approximately given by the location over time where the (non-stationary) cross-correlation function takes on its global peak. Motivated by the asymptotic result described in the first part of this chapter, a Poisson model for the large errors is applied to derive an approximate expression for the global variance of the correlator estimate. This expression is shown to be an approximate upper bound on the variance, which complements previously derived lower bounds for the time delay estimation problem, e.g. Cramer-Rao [10], Barankin [2], Ziv-Zakai [20] and others [17].

Section II and III present the relevant theory concerning point process representations and asymptotic theory. Section IV contains a discussion of the application of the Poisson model to peak ambiguity in two-sensor arrays. Finally in Section V we compare the Poisson variance approximation to the Ziv-Zakai-Lower-Bound (ZZLB), the Cramer-Rao-Lower-Bound (CRLB) and experimental results.

## II: A REPRESENTATION FOR THE PROBABILITY OF LEVEL CROSSINGS

Let $(\Lambda, \Phi, P)$ be a complete probability space and define the nested sequence of $\sigma$-fields : $\Phi_t$, $t \in R$, $(\Phi_t \subset \Phi,\ \Phi_s \subset \Phi_u$ for $s < u)$. We assume the $\Phi_t$-measurable random process $X(t)$, $-\infty < t < +\infty$, to have the following properties: separability, almost sure (a.s.) sample function continuity, existence of the bivariate densities, $f_{t,\tau}(y,z)$, of $X(t)$ and $X(\tau)$, for $t \neq \tau$. Additional properties will be imposed shortly. For expository convenience we focus on the relevant level crossing theory for upcrossings. However, the consideration of downcrossings is completely analogous to the presentation to follow.

We define an upcrossing analogously to Leadbetter in [11]. A realization of $X(t)$, $x(t)$, upcrosses zero in the interval $[\sigma,\nu)$ if there exists an open interval centered at some point

$t_u \in (\sigma,\nu)$, $(t_u - \delta, t_u + \delta)$ say, over which $X(t) < 0$ to the left of $t_u$ and $X(t) > 0$ to the right of $t_u$. We denote this occurrence by the notation $A_{\sigma,\nu}$. Symbolically

$$A_{\sigma,\nu} \equiv \{w \in \Lambda | \exists\ t_u \in (\sigma, \nu), \exists\ \delta > 0\ ;\ s.t.\ X_{t_u - h} < 0 < X_{t_u + h},\ \text{for } 0 < h < \delta\} \tag{1}$$

This definition essentially excludes any "non-regular" upcrossings such as tangencies or non-smooth intersections of zero. It is shown in [11] that $A_{\sigma,\nu}$ is $\Phi_\nu$-measurable and that non-regular upcrossings are of zero probability. Hence definition (1) is sufficiently general for our purposes. The definition of downcrossings is analogous to (1).

We further define the number of upcrossings by $X(\tau)$ in $[\sigma, \nu)$, denoted $N(\sigma, \nu)$, as the number of distinct points $t_u$ at which upcrossings occur. More specifically we define: $N(\sigma,\nu) = \limsup_{\Delta \to \infty} \sum_{t_i} I(A_{t_i, t_i + \Delta})$. Here $I(A)$ is the indicator of the event $t_u \in A$ and $\{t_i\}$ is an increasingly dense partition of $(\sigma,\nu)$ with inter grid spacing $\Delta$.

By the assumed continuity of the random process $X(t)$ it is reasonable to expect that it can be well approximated by a piecewise linear process tied to $X(t)$ at a sufficiently dense set of points $t = t_o, t_1, \ldots, t_M$. That is, let $\xi_n(t)$ denote a random process defined on an interval $[t_o, t_f]$ for which

$$\xi_n(t) = \begin{cases} X(t) & t = t_k \\ X(t_k) + \dfrac{[X(t_{k+1}) - X(t_k)]}{(t_f - t_o)2^{-n}}(t - t_k) & t_k < t < t_{k+1} \end{cases} \tag{2}$$

$$t_k = t_o + k\,2^{-n}(t_f - t_o),\ k = 0, 1, 2, \cdots, 2^n$$

If $N_n$ is the number of upcrossings of zero by $\xi_n(t)$ then the following is due to Ylvisaker [21].

*Lemma 1.1*

*Let $N_n(t)$ be the number of upcrossings of zero by $\xi_n$ in the interval $[t_o, t)$. Then $N_n(t)$ is monotonically non-decreasing in n and converges to $N(t)$, the number of upcrossings of zero by $X$ in the same interval, with probability one as $n \to \infty$.*

From the above lemma it follows by monotone convergence that

$$P(N(t) \le k) = \lim_{n \to \infty} P(N_n(t) \le k),\ \text{for } k = 0,1,2,\cdots \tag{3}$$

Hence, as far as the computation of upcrossing probabilities is concerned, $\xi_n(t)$ and $X(t)$ can be used interchangeably in the sense of (3).

The following will be important for the upcoming development and are essentially Theorems (2.1) and (2.2) of [13].

*Lemma 1.2*

*Let $[t_o, t_f\ ]$ have the partition $\{t_i\}_{i=0}^{2^n}$. Then with $N(t)$ the number of upcrossings (downcrossings) of zero by $X$ in $[t_o, t)$ and $N_n(\tau, \sigma)$ the number of upcrossings (downcrossings) by $\xi_n$ in $[\tau, \sigma)$*

$$\mathbf{E}\,[N(t_f)] = \lim_{n \to \infty} \sum_{i=0}^{2^n-1} P(A_{t_i, t_{i+1}}) = \lim_{n \to \infty} \sum_{i=0}^{2^n-1} P(N_n(t_i, t_{i+1}) > 0) \tag{4}$$

Define $g_{t,\tau}(y,z)$ the joint density of $X(t)$ and $[X(t+\tau) - X(t)]/\tau$. Then by elementary transformations

$$g_{t,\tau}(y,z) = \tau f_{t,t+\tau}(y, y+\tau z) \tag{5}$$

The following are essential to the development and are known as Leadbetter's conditions [11, Thm. 2]

$$g_{t,\tau}(y,z) \quad \textit{is continuous in } t, y \text{ for } \textit{each } \tau, z \tag{6}$$

$$g_{t,\tau}(y,z) \to \mathrm{p}_t(y,z) \quad \textit{as } \tau \to 0 \textit{ uniformly in } t, y \tag{7}$$

$$g_{t,\tau}(y,z) \leq l(z) \quad \text{for } \textit{all } t, \tau, y, z \tag{8}$$

*where*

$$\int_{-\infty}^{\infty} |z| \; l(z) dz < \infty$$

If the above three conditions are satisfied then the following representation theorem holds for the probability of getting an upcrossing in $[t_o, t)$, $P(A_{t_o, t})$, here denoted $p(t)$.

*Theorem 1.1*

*Suppose $X(\tau)$ has continuous sample functions with probability one and let the conditions (6) through (8) hold. Then the expected value of the number of upcrossings of zero by $X(\tau)$ in any finite interval $[t_o, t_f\ )$ is finite and given by*

$$\mathbf{E}\,[N(t_o, t_f)] = \int_{t_o}^{t_f} \rho(\tau) d\tau \tag{9}$$

$$\rho(\tau) = \int_0^{\infty} z \ p_\tau(0,z)dz$$

*Furthermore the probability of getting at least one upcrossing of zero by $X(\tau)$ in $[t_o, t)$, $p(t)$, satisfies the relation*

$$p(t) = \int_{t_o}^{t} \rho(\tau)(1 - p(\tau))d\tau + Q(t) \tag{10}$$

*where*

$$Q(t) = \lim_{n \to \infty} \sum_{i=0}^{2^n - 1} q(t_i) \tag{11}$$

$$q(t_i) = P(N_n(t_i, t_{i+1})>0, N_n(t_i)=0) - P(N_n(t_i, t_{i+1})>0)P(N_n(t_i)=0) \tag{12}$$

*Here $\{t_i\}_{i=0}^{2^n}$ is a partition of $[t_o, t]$*

The probability of downcrossings satisfies an analogous Theorem, the only difference being the particular form of the intensity $\rho$ in (9). For downcrossings $\rho(\tau) \triangleq \int_{-\infty}^{0} |z| p_\tau(0,z)dz$.

The function $\rho(\tau)$, given as the derivative of (9), is the incremental average number of level crossings per unit time at time $\tau$, $\rho(\tau)d\tau = \mathbf{E}\{dN(\tau)\}$. In the theory of point processes $\rho$ is called the (incomplete) intensity function of the point process $N$. $Q(\tau)$ in (12) can be interpreted as a measure of the dependency structure of the upcrossing process $N$ over disjoint intervals (for independent increment processes $Q$ would be zero).

Eq. (9) of Theorem 1.1 is obtained directly by modifying the proof of Theorem 2 of Leadbetter for downcrossings [11] to the case of upcrossings. The proof of the rest of Theorem 1.1 depends on a particular decomposition of the event that an upcrossing of zero by $\xi_n$ occurs on $[t_o, t)$, which we denote $B_{t_o,t}$. If $N_n(t)$ is finite we can define $B^1_{\sigma,\nu}$: the event that the first instance of an upcrossing occurs in the subinterval $[\sigma,\nu)$ of $[t_o, t)$. That is

$$B^1_{\sigma,\nu} \equiv B_{\sigma,\nu} \cap \overline{B}_{t_o,\sigma} \tag{13}$$

where we read this as: $\xi_n$ first upcrosses in $[\sigma,\nu)$ if there is an upcrossing in $[\sigma,\nu)$ but none in $[t_o, \sigma)$.

We note the following two rather obvious properties of $B^1_{\sigma,\nu}$.

For $[\sigma,\nu)$ and $[s, t)$ disjoint

$$B^1_{\sigma,\nu} \text{ and } B^1_{s,t} \text{ are disjoint} \tag{14}$$

and

$$B^1_{t_o,t} \equiv B_{t_o,t} \quad , t \in [t_o,t) \tag{15}$$

Eqs. (14) and (15) follow directly from the definition (13). The following proposition is central to the decomposition alluded to above.

*Proposition 1.1*

*Let $A_{\sigma,\nu}$ denote an upcrossing of zero, and $A^1_{\sigma,\nu}$ the first instance of an upcrossing, by a random process $X(t)$ in $[\sigma,\nu)$, where $X(t)$ has absolutely continuous distributions. Then if the expected number of upcrossings of zero in $[t_o,t)$, $\mathbf{E}\{N(t_o,t)\}$, is finite the following equivalence holds with probability one*

$$A_{t_o,t} \equiv \bigcup_{i=0}^{2^n-1} A^1_{t_i,t_i+1} \tag{16}$$

*where $\{t_i\}_{i=0}^{2^n}$ is a partition of $[t_o,t]$*

∎

*Proof*

Note that the number of upcrossings in $[t_o,t)$ is finite with probability one since

$$P(N(t_o,t) > k) \leq \sum_{i=k+1}^{\infty} P(N(t_o,t) = i) \tag{17}$$

$$\leq \sum_{i=k+1}^{\infty} iP(N(t_o,t) = i)$$

which must converge to zero as $k \to \infty$ by the finiteness of the mean number of upcrossings. Thus a "first instance of an upcrossing" is well defined. The inclusion "$\supset$" in (16) is trivial since any upcrossing in a subinterval of $[t_o,t)$ implies an upcrossing occurred in the entire interval. As for "$\subset$", if there is an upcrossing in $[t_o,t)$ it is either interior to one of the $[t_i,t_{i+1})$ or at one of the endpoints $t_i$ , $i = 0, 1, \ldots, 2^n-1$ However from the absolute continuity of the distribution of $X(t)$, with respect to Lebesgue measure, this latter event has probability zero. Therefore the proposition follows.

∎

*Proof of Theorem 1.1*

Partition $[t_o,t)$ into $2^n-1$ subintervals of length $\Delta = (t_f - t_o)2^{-n}$ for $n = 0,1, \cdots$. That is we have intervals $[t_i,t_{i+1})$ with $t_i = t_o + i\Delta$ , $i = 0,1, \ldots, 2^n$. Define $B_{\sigma,\nu}$,

$\sigma,\nu \in \{t_i\}_{i=1}^{2^n}$, the event that the polygonal approximation, $\xi_n$ upcrosses zero in $[\sigma,\nu)$, i.e $N_n(\sigma,\nu) > 0$. Then from Proposition 1 and Eq. (15) for $k = 2^n$ ($t_k = t$)

$$P(B_{t_0,t_k}) = P(\bigcup_{i=0}^{2^n-1} B_{t_i,t_{i+1}} \cap \overline{B}_{t_0,t_i}) = \sum_{i=0}^{2^n-1} P(B_{t_i,t_{i+1}} \cap \overline{B}_{t_0,t_i}) \tag{18}$$

Here we have used the disjointness property (14). Now add and subtract the product $P(B_{t_i,t_{i+1}})P(\overline{B}_{t_0,t_i})$ from each term under the sum (18)

$$P(B_{t_0,t}) = \sum_{i=0}^{2^n-1} [P(B_{t_i,t_{i+1}})P(\overline{B}_{t_0,t_i}) + q(t_i)] \tag{19}$$

where

$$q(t_i) = P(B_{t_i,t_{i+1}} \cap \overline{B}_{t_0,t_i}) - P(B_{t_i,t_{i+1}})P(\overline{B}_{t_0,t_i}) \tag{20}$$

as in the statement of the Theorem, Eq. (12).

$B_{t_i,t_{i+1}}$ is equivalent to the event

$$B_{t_i,t_{i+1}} \equiv \{\xi_n(t_i) < 0 < \xi_n(t_{i+1})\} \tag{21}$$

Define

$$\eta_n(t_i) = \frac{\xi_n(t_{i+1}) - \xi_n(t_i)}{\Delta} \tag{22}$$

Combining Eqs. (21) and (22)

$$B_{t_i,t_{i+1}} = \{\xi_n(t_i) < 0 < \xi_n(t_i) + \Delta\eta_n(t_i)\} \tag{23}$$

$$= \{\xi_n(t_i) \in (-\Delta z, 0)\ ,\ \eta_n(t_i) = z > 0\}$$

Therefore by the definition of the joint density, $g_{t_i,\delta}$, of $\xi_n(t_i)$ and $\eta_n(t_i)$

$$P(B_{t_i,t_{i+1}}) = \int_0^\infty dz \int_{-\Delta z}^0 g_{t_i,\Delta}(x,z)\,dx \tag{24}$$

Now make a change of variable in the argument $x$ of (24) and substitute the result back into Eq. (19) to obtain

$$P(B_{t_0,t}) = \sum_{i=0}^{2^n-1} [\Delta \int_0^\infty dz \int_{-z}^0 g_{t_i,\Delta}(\Delta\, x, z) P(\overline{B}_{t_0,t_i})\,dx + q(t_i)] \tag{25}$$

By the pointwise continuity and uniform convergence conditions, (6) and (7), for $\Delta$ sufficiently small

$$\int_{-z}^{0} g_{t_i,\Delta}(\Delta x, z)\,dx = z\ \mathrm{p}_\tau(0,z)\ , \qquad \tau \in [t_i, t_{i+1}] \tag{26}$$

Condition (8) asserts that $zl(z)$ is integrable over the positive real line where $l(z)$ upper bounds $g_{t,\tau}(y,z)$. Therefore the limit of (26) as $\Delta \to 0$ is bounded except possibly on some set of measure zero. From Lemma 1.1 and Eq. (3) $P(\overline{B}_{t_o,t_i})$ converges to $P(\overline{A}_{t_o,t_i}) = 1 - p(t_i)$. Defining

$$a_\Delta(t_i, z) = \int_{-z}^{0} g_{t_i,\Delta}(\Delta x, z) P(\overline{B}_{t_o,t_i})\,dx \tag{27}$$

we have as $\Delta$ goes to zero

$$a_\Delta(t_i, z) \to z\ \mathrm{p}_{t_i}(0,z)(1 - p(t_i)) \qquad a.e. \tag{28}$$

and

$$a_\Delta(t_i, z) \le zl(z) \tag{29}$$

Hence by dominated convergence the first term of the expression (19) becomes in the limit

$$\lim_{n\to\infty} \sum_{i=0}^{2^n-1} P(B_{t_i,t_{i+1}}) P(\overline{B}_{t_o,t_i}) \tag{30}$$

$$= \lim_{n\to\infty} \Delta \sum_{i=0}^{2^n-1} \int_0^\infty dz \int_{-z}^{0} g_{t_i,\Delta}(\Delta\, x, z) P(\overline{B}_{t_o,t_i})$$

$$= \lim_{n\to\infty} \int_{t_o}^{t} d\tau \int_0^\infty dz \int_{-z}^{0} dx\ \ g_{\tau,\Delta}(\Delta\, x, z) P(\overline{B}_{t_o,\tau})$$

$$= \int_{t_o}^{t} d\tau \int_0^\infty z\ \mathrm{p}_\tau(0,z)(1 - p(\tau))\,dz < \infty$$

This is the first additive term in Eq. (10).

From the expression (20)

$$-P(B_{t_i,t_{i+1}}) P(\overline{B}_{t_o,t_i}) \le q(t_i) \le P(B_{t_i,t_{i+1}}) P(B_{t_o,t_i}) \tag{31}$$

so that the $q(t_i)$ are absolutely summable by Lemma 1.2 and the finiteness of $\mathbf{E}[N(t_o, t_f)]$. Finally Theorem 1.1 follows by performing the limiting operation in (16) as $n \to \infty$, taking account of the regularity conditions shown above. ∎

## III. ASYMPTOTIC RESULTS

Theorem 1.1 gives an implicit relation for the probability of getting an upcrossing in a bounded interval. Although the intensity function $\rho$ may be known, in general the $Q(t)$ term in Eq. (11) involves quantities which are not known. On the other hand Eq. (11) can be used to prove certain asymptotic results for a fairly general class of upcrossing processes, which we will now undertake to show. To motivate these results the following argument is useful. Referring to Eq. (11), assume that $N^m$ converges to an independent increment point process $\widetilde{N}$ as $m \to \infty$. Then $q^m(t_i)$ converges to zero for all $i$ and by Eq. (31), Lemma 1.2 and the finiteness of the mean number of upcrossings, $q^m(t_i)$ is summable over $i$ as the $t_i$ become dense in $[t_o, t)$. Dominated convergence then assures that $Q(t) = 0$ and Eq. (11) becomes equivalent to a linear first order, homogeneous differential equation with coefficient $\rho(\tau)$ and initial condition $p(t_o) = 0$. Eq. (11) then has the solution

$$p(t) = 1 - exp\left(-\int_{t_o}^{t} \rho(\tau) d\tau\right) \tag{32}$$

Eq. (32) is of course valid for any semiclosed subinterval of $[t_o, t)$. Hence, by the independence of $N$ over disjoint intervals, the upcrossing process must actually be an inhomogeneous Poisson process with intensity $\rho$.

Unfortunately the above argument is fallacious since, roughly speaking, for non-zero $N$ on bounded intervals, the independent increment property of $N$ is incompatible with the sample function continuity of $X$ so that Theorem 1.1 does not even apply. Clearly the pointwise convergence of $N^m$ to an independent increment process $\widetilde{N}$ is an overly strong imposition on $X$. However in the following it will be shown that for a sequence of "thinned out" upcrossing count processes, $N^m(t_o, t)$ associated with $X$, a related (time normalized) random counting process can be defined which converges in distribution to a Poisson random process defined on the interval $[0,1)$ as $m \to \infty$. These results will depend on additional assumptions, such as mixing, on the distributions of $X$.

The basic idea is as follows. Let $X_0 \triangleq \{X(t): t \in [0,1]\}$ be a given random process with upcrossing intensity $\{\rho(t): t \in [0,1]\}$. Define a sequence of increasingly long time intervals, $I_m$, of length $T_m > 1$, $I_m \triangleq [0, T_m]$. On the interval $I_m$ let $X_m$ be a random process with an upcrossing intensity function $\{\rho^m(t): t \in [0, T_m]\}$ where $\rho^m$ is related to $\rho$ by: $\rho^m(t) \triangleq \frac{1}{T_m} \rho(\frac{t}{T_m})$, $t \in [0, T_m]$. Note, the average number of upcrossings by $\{X(t): t \in [0,1]\}$ is over $[0,1]$ is identical to the average number by $\{X_m(t): t \in [0, T_m]\}$ over $[0, T_m]$, while the intensity $\rho^m$ is a stretched and downscaled (thinned) version of $\rho$. In this way the upcrossings by $X_m$ differ from those of $X$ only in that the average inter-event spacing has been uniformly increased, i.e. upcrossings by $X_m$ become "rare events" over time. As we increase $T_m$ out to infinity, the upcrossings will become approximately independent since, with probability close to one, the events are separated in time by an amount exceeding the "interdependence time" (correlation time for Gaussian case ) of $X_m$ which can be specified by a mixing condition. Then, with the aid of some additional regularity conditions, Theorem 1.1 can be used to give the solution (32).

For simplicity, and without restricting the generality of the results, we set $t_o$ in Eq. (32) to zero. In general, when there is multiple indexing, subscripts indicate indexing with respect to the partition, $\{t_i\}$, of the time interval under consideration and superscripts index the quantity with respect to the infinite sequences $\{X_m\}$ and $\{I_m\}$. Thus $N_n^m(t)$ denotes the number of upcrossings of zero by the polygonal approximation to $X_m$, $\xi_n^m$, over the interval $[0,t)$, $t \in I_m$. Likewise $N^m$ is the number of upcrossings associated with $X_m$ itself. Analogously to the development of Theorem 1.1 define $\Phi_{\sigma,\nu}^m$, the $\sigma$-field generated by $X_m$ on $[\sigma,\nu)$; $B_{t_i,t_j}^m$, the event $N_n^m(t_i,t_j) > 0$, where $t_i$ and $t_j$ are points contained in the $2^n$-th order partition of $I_m$; and $p_m(t)$, the probability that $X_m$ upcrosses zero on $[0,t) \subseteq [0,T_m)$.

Throughout the sequel of this section, we assume the intensity associated with $N^m$, $\rho^m$ exists for all $m$ and is defined in terms of the intensity associated with $N^0$, $\rho$, as follows

$$\rho^m(\tau) \triangleq \frac{1}{T_m} \rho(\frac{\tau}{T_m}), \quad m = 0, 1, \cdots \tag{33}$$

The next section is concerned with the various conditions which will be imposed on $X_m$ for asymptotic independence of widely separated segments of the trajectory and Poisson-like behavior of the upcrossings. While not necessarily the most compact set of sufficient conditions, the following contribute to a clear and simple proof of the asymptotic theorem. Several comments will be made concerning simpler sufficient conditions during the discussion.

Asymptotic Conditions:

A mixing condition is a statement concerning the asymptotic independence of the trajectories of a random process on disjoint intervals $[\sigma,\nu)$ and $[s,\tau)$ as $|s - \nu| \to \infty$. For example $X$ is "strong mixing" [16] if

$$\sup_{\tau} |P(A \cap B) - P(A)P(B)| < \beta_l \tag{34}$$

where $A$ and $B$ are arbitrary events

$$A \in \Phi_{\tau,\infty}, \quad B \in \Phi_{-\infty,\tau-l}$$

and

$$\lim_{l \to \infty} \beta_l = 0$$

The major weakness of "strong mixing" is that (34) becomes vacuous if either $A$ or $B$ are of vanishingly small probability. Indeed in the present context the event $A$ will be contained in the event that an upcrossing of zero occurs in an exceedingly small interval, which of course has exceedingly small probability. The needed condition here is the summability to zero of the differences below

$$\lim_{m,n \to \infty} \sum_{i=0}^{2^n-1} |P(A_i^m \cap B_i^m) - P(A_i^m)P(B_i^m)| = 0 \tag{35}$$

$$A_i^m \in \Phi^m_{-\infty, t_i - l_m} , \quad B_i^m \in \Phi^m_{t_i, t_{i+1}}$$

$$l_m \to \infty , \; l_m = o(T_m) \; \text{as } n, m \to \infty$$

$$\{t_i\}_{i=0}^{2^n} , \quad \text{an increasingly dense partition of } [0, T_m)$$

A sufficient condition for (35), if the quantities $P(B_i^m)$ are summable over $i$, is the following form of so called "uniform mixing" [16].

*Mixing Condition*

*With $\Phi^m_{\sigma,\nu}$ the $\sigma$-field generated by the trajectories of $X_m$ in $[\sigma,\nu)$, $T_m$ a monotonic sequence increasing to infinity, $X_m$ is said to be uniform-asymptotically mixing (u-a mixing) if*

$$|P(A_i^m) - P(A_i^m | B_i^m)| < \alpha_{m, l_m} \tag{36}$$

*where*

$$A_i^m \in \Phi^m_{-T_m, t_i - l_m} , \quad B_i^m \in \Phi^m_{t_i, t_{i+1}}$$

*and*

$$\lim_{m \to \infty} \alpha_{m, l_m} = 0$$

*with*

$$l_m \to \infty, \; l_m = o(T_m) , \; \text{as } n, m \to \infty$$

$$\{t_i\}_{i=0}^{2^n} , \quad \text{an increasingly dense partition of } [0, T_m)$$

Note that for a dense partition $\{t_i\}$ the conditioning in (36) will be on the zero probability event $X(\tau) = 0$ at some specific point $\tau$, viewed as a limit through a horizontal window [8]. Thus in the limit of dense partitions, although the conditional probability may not be well defined in the conventional sense, (36) is well defined. We state the following lemma [5] which generalizes the uniform mixing condition to multiple events.

*Lemma 2.1*

*Assume that $X_m(t)$ is uniform mixing in the sense of (36). Fix $l > 0$ and for $r > 1$ let $E_1, E_2, \ldots, E_r$ be disjoint intervals indexed in increasing order, that is, $\sup\{\tau \in E_{i-1}\} < \inf\{\tau \in E_i\}$ for $i = 1, 2, \ldots, r$, and separated by at least $l$. Then* *for* $A_i^m \subset \Phi_{E_i}^m$

$$|P(\bigcap_{i=1}^{r} A_i^m) - \prod_{i=1}^{r} P(A_i^m)| < \alpha_{m,l_m} \sum_{i=2}^{r} P(A_i^m) \tag{37}$$

For Gaussian processes $X_m$ a sufficient condition for mixing is a rate of decay on its auto-correlation function: $R_{X_m}(t,t+l_m)\log l_m = \alpha_m(t) \to 0$ as $m, l_m \to \infty$ uniformly in $t$ [14]. In order to make the upcrossings exceedingly rare events as $m \to \infty$ the following "rarefaction" condition is used

*Rarefaction Condition*

*With* $N_n^m(\sigma,\nu)$ *the upcrossings of zero by the polygonal approximation* $\xi_n^m$ *in* $[\sigma,\nu) \subset [0,T_m)$ $N_n^m$ *satisfies a rarefaction condition if for* $l_m \to \infty$, $l_m = o(T_m)$

$$\lim_{n,m\to\infty} \sum_{t_i=l_m}^{T_m} P(N_n^m(t_i,t_{i+1}) > 0,\ N_n^m(t_i-l_m,t_i) > 0) = 0 \tag{38}$$

$$\{t_i\}_{i=0}^{2^n}, \quad \textit{an increasingly dense partition of } [0,T_m)$$

The above condition is a strong condition on the trajectories similar to, but possibly more restrictive than, the condition $D_c'$ used in [12] for the stationary case. Eqn. (38) guarantees that the probability of more than a single level crossing over any $o(T_m)$ interval be exceedingly small as $m \to \infty$. The condition (38) is somewhat stronger than the property of *a-regularity* for $a = 2$ (see Lemma 2.3). (35) can be shown to hold if the hazard function,

$$h^m(u,\tau) \triangleq \lim_{h\to 0}\frac{1}{h} P(N^m(u,u+\tau) = 0 \mid N^m(u-h,u) > 0), \quad \tau \geq 0 \tag{39}$$

satisfies $1 - h^m(u,l_m) = o(\frac{1}{T_m})$ for all $u \in [0,T_m-l_m]$.

An additional condition needed is the following which is analogous to condition (4.6) in [12]

$$\frac{P(N_n^m(t,t+h) > 0)}{\mathbf{E}[N_n^m(t,t+h)]} \to 1 \quad \textit{as } n,m \to \infty \tag{40}$$

$$\{t_i\}_{i=0}^{2^n}, \quad n = n(m), \quad \textit{an increasingly dense partition of } [0,T_m)$$

for *some* $h_o$, $0 < h < h_o$ and for *all* $t \in [0,T_m)$.

Condition (40) is stronger than a well known necessary condition for a process to be (asymptotically) Poisson: for infinitesimal intervals the probability of getting a point is proportional to the expected number of points in the interval (linear in the length of the interval for stationary processes). The condition can be interpreted as an extension of this necessary condition to certain finite intervals.

We state here two easily verifiable conditions on the average number of crossings by $X_0$ over $I_0 = [0,1)$, $\mathbf{E}\{N(1)\}$, which are particular to the nonstationary situation. These properties guarantee that the behavior of the upcrossing process $N^m$ be sufficiently uniform over time to exclude degeneration of the upcrossing probabilities to either probability 0 or probability 1 events over any $o(T_m)$ interval.

*Uniform Denseness Condition*

*Let $N$ be the number of upcrossings by $X_0$ on $I_0 = [0,1)$. Choose an interval $A$, a subset of $[0,1)$. The uniform denseness condition is satisfied if for any $\epsilon, K$, $\epsilon > 0$, $1 < K < \infty$, there exist $K$ subintervals of $A$, $\{J_i\}_{i=1}^{i=K}$, whose closures are disjoint, such that*

$$|\mathbf{E}[N(J_i)] - \mathbf{E}[N(J_l)]| < \epsilon, \qquad i \neq l,\ i,l = 1, \ldots, K \tag{41}$$

*Asymptotic Uniform Negligibility*

*Let $N$ be as in the condition above and let $\{\tau_i\}_{i=0}^{K}$ be a uniformly spaced partition of $I_0 = [0,1)$ of size $K$. Then with $N_{\tau_k} = N(\tau_k, \tau_{k+1})$, the number of upcrossings within the $k$-th partition element, the uniform negligibility condition is satisfied if for all $l = 1, \ldots, K$*

$$\lim_{K \to \infty} \frac{\mathbf{E}[N_{\tau_l}]}{\sum_{k=1}^{K} \mathbf{E}[N_{\tau_k}]} = 0 \tag{42}$$

Loosely speaking (41) implies that the upcrossings are lean enough so that "similar" intervals, of similar order with respect to $I_0$, have associated with them a "similar" expected number of upcrossings. This will imply a continuity property on $P(N^m(\tau,\sigma) > 0)$ viewed as a function from the sets $[\tau,\sigma)$. Condition (42) guarantees that in no case will the total number of upcrossings over $I_0$ be dominated by upcrossings in small subintervals of $I_0$. The reader may be interested in the similarity between Asymptotic Uniform Negligibility and Feller's sufficient condition for a non-stationary Central Limit Theorem [22]. If the process $X_0$ were stationary, these two conditions, (41) and (42), would be trivially satisfied since the expected values of $N(J)$ and $N(I)$ are identical if $J$ and $I$ are intervals of equal length. For non-stationary $X_0$, a sufficient condition for (41) and (42) is that the (incomplete) intensity, $\rho = \rho_0$, satisfy $0 < \rho < M$ for some finite $M$.

Main Theorem

With the above conditions we are prepared to state the main result concerning the convergence of a certain normalized upcrossing process, associated with $X_m(t)$, to a Poisson process.

*Theorem 2.1*

*Let the a.s. continuous processes $X_m(t)$ have absolutely continuous distributions for $m = 1, \cdots$. Assume each $X_m$ satisfies Leadbetter's conditions (6) through (8), with, in addition, $l_m(z) = O(T_m^{-1})$ in (8), u-a mixing of the form (36) and conditions surrounding Eqs. (38) through (42). Also assume that the (incomplete) intensity, $\rho^m$, of the zero upcrossings by $X_m$ over $[0,\tau)$, $\tau \in [0, T_m]$, $N^m(\tau)$, satisfy (33). Then if the time normalized count process $\widetilde{N}^m$ is defined: $\widetilde{N}^m(\tau) \triangleq N^m(\tau T_m)$, $\tau \in [0,1]$, we have*

$$\widetilde{N}^m(\tau) \rightarrow N^*(\tau) \text{ in distribution} \tag{43}$$

*where $N^*(\tau)$ is a non-stationary Poisson random process on $[0,1)$ with intensity $\rho$.*

We will need the following proposition in order to use Theorem 1.1 for the proof of Theorem 2.1.

*Proposition 2.1*

*Let $X_m(t)$ be a random process which satisfies the conditions in Theorem 2.1. Let $N^m(\sigma,\tau)$ and $N_n^m(\sigma,\tau)$ be the number of upcrossings of zero within $[\sigma,\tau)$ by $X_m$ and the approximation to $X_m$, $\xi_n^m$, respectively. Further assume that $T_m \rho^m(\tau T_m) = \rho(\tau)$. Then for $p_m(t)$ the probability that $\widetilde{N}^m(0,t) \triangleq N^m(0,tT_m)$ exceeds zero and $p^*(t)$ the probability that a Poisson count process with intensity function $\rho$ exceeds zero in the interval $[0,t) \in [0,1]$.*

$$| p_m(t) - p^*(t)| \leq \lim_{n \rightarrow \infty} \sum_{i=0}^{2^n-1} |q^m(t_i)| \, exp(-\int_0^t \rho(\tau) d\tau) \tag{44}$$

*where*

$$q^m(t_i) = P(B_{t_i,t_{i+1}}^m, \overline{B}_{0,t_i}^m) - P(B_{t_i,t_{i+1}}^m) P(\overline{B}_{0,t_i}^m) \tag{45}$$

$B_{t_i,t_j}^m \triangleq \{N_n^m(t_i,t_j) > 0\}$. *for an increasingly dense partition, $\{t_i\}_{i=0}^{2^n-1}$ of the interval* $[0, T_m]$.

*Proof*

Since $X_m$ satisfies the assumptions of Thm. 1.1 and $P(N^m(tT_m) > 0) = p_m(t)$, $t \in [0, T_m]$

$$p_m(t) = \int_0^{tT_m} \rho_m(\tau)(1 - p_m(\tau/T_m))d\tau + \lim_{n\to\infty}\sum_{i=1}^{2^n-1} q^m(t_i), \quad \tau\in[0,1] \tag{46}$$

With a change of variable $\tau/T_m \to \tau$, and the identity relating $\rho$ to $\rho_m$ (33)

$$p_m(t) = \int_0^{t} \rho(\tau)(1 - p_m(\tau))d\tau + \lim_{n\to\infty}\sum_{i=1}^{2^n-1} q^m(t_i) \tag{47}$$

$p^*$ is a Poisson probability measure having the form $p^*(t) = exp(-\int_0^t\rho(\tau)d\tau)$ hence $p^*$ satisfies the integral equation

$$p^*(t) = \int_0^t\rho(\tau)(1 - p^*(\tau))d\tau \tag{48}$$

An application of the triangle inequality to the difference: (47) minus (48), yields

$$| p_m(t) - p^*(t)| \le \int_0^t\rho(\tau)| p(\tau) - p^*(\tau)|d\tau + \lim_{n\to\infty}\sum_{i=1}^{2^n-1} | q^m(t_i)| \quad , \quad t' \in [0,t] \tag{49}$$

Let the last term in (49) be denoted $r(tT_m)$. Then $r$ is monotone non-decreasing in $t \in [0,1]$ and replacement of $r(tT_m)$ by $r(T_m)$ can only weaken the inequality (49). Subsequent application of the Bellman-Gronwall inequality [18] to (49) finishes the proof.

∎

The following generalization of Lemma 2.2.3 in [12] is proven in [5].

*Lemma 2.2*

*Let $X_m$ satisfy (6) through (8), be u-a mixing and satisfy (38) and (40). Then given $\epsilon > 0$, integers $r > 0$, $K > 0$, positive quantities $l$, $l = o(T_m/K)$ and $t$, $\frac{T_m}{K} < t < T_m$, we have for m sufficiently large*

$$P(N_n^m(t-l,t) > 0,\ N_n^m(t-l) = 0) < \left(\frac{r}{r+1}\right)\frac{1}{r+1} + (2r-1)\alpha_{m,l} + \epsilon \tag{50}$$

With Lemma 2.2 and Proposition 2.1 we can easily prove the following weak form of the Poisson convergence result.

*Proposition 2.2*

*If the a.s. continuous processes $\{X_m(t)\}$ satisfy the conditions stated in the premise of Theorem 2.1 then for any interval I contained in [0,1]*

$$P(\tilde{N}^m(I) > 0) \to 1 - exp(-\int_I\rho(\tau)d\tau) \tag{51}$$

*Proof*

First fix $l$ greater than zero and $K$ a positive integer. We reproduce Eq. (45) here for clarity. As in Eq. (12) of Theorem 1.1, for the sequence $X_m$, $m = 0, 1, \cdots$ we have the quantities $q^m(t_i)$ on the $2^n$ point grid $\{t_i\}_{i=0}^{2^n}$

$$q^m(t_i) = P(B^m_{t_i, t_i+1}, \bar{B}^m_{0,t_i}) - P(B^m_{t_i,t_i+1})P(\bar{B}^m_{0,t_i}) \tag{52}$$

Partition the interval $[0, T_m)$ into $K$ parts so that the sum on the right of Eq. (44) of Proposition 2.1 can be represented as

$$\sum_{i=0}^{2^n-1} |q^m(t_i)| = \sum_1 |q^m(t_i)| + \cdots + \sum_K |q^m(t_i)| \tag{53}$$

where $\sum_k$ denotes summation over the intersection of the grid $\{t_i\}_{i=0}^{2^n}$ and the $k$-th partition element, $k = 1, 2, \ldots, K$.

Fix $\epsilon > 0$ and let $m$ be sufficiently large so that Lemma 2.2 holds. Consider the final $K-1$ terms in (53)

$$\sum_2 |q^m(t_i)| + \cdots + \sum_K |q^m(t_i)| \tag{54}$$

Now for each $q^m(t_i)$ in (54) we add and subtract terms so as to isolate the mixing dominated quantities of the form (36). That is we obtain via the triangle inequality

$$\begin{aligned} |q(t_i)| \le\ & |P(B_{t_i,t_i+1}, \bar{B}_{0,t_i-l}) - P(B_{t_i,t_i+1})P(\bar{B}_{0,t_i-l})| \\ & + |P(B_{t_i,t_i+1})P(\bar{B}_{0,t_i-l}) - P(B_{t_i,t_i+1})P(\bar{B}_{0,t_i})| \\ & + |P(B_{t_i,t_i+1}, \bar{B}_{0,t_i}) - P(B_{t_i,t_i+1}, \bar{B}_{0,t_i-l})| \end{aligned} \tag{55}$$

where we have suppressed dependencies on $m$ for notational simplicity. Using the mixing condition (36) on the first term to the right of the inequality (55) and using simple set identities for the other two terms we have

$$|q(t_i)| \le P(B_{t_i,t_i+1})\,[\alpha_{m,l} + P(B_{t_i-l,t_i}, \bar{B}_{0,t_i-l})] + P(B_{t_i,t_i+1}, B_{t_i-l,t_i}, \bar{B}_{0,t_i-l}) \tag{56}$$

Finally applying Lemma 2.2 to the second term in brackets [ ] in (56) and noting that by monotonicity the third term in (56) is bounded

$$P(B^m_{t_i,t_i+1}, B^m_{t_i-l,t_i}, \bar{B}^m_{0,t_i-l}) \le P(N^m_n(t_i, t_{i+1}) > 0,\ N^m_n(t_i-l, t_i) > 0) \tag{57}$$

we get by substituting the inequality (56) in (53)

$$\sum_{i=0}^{2^n-1} |q^m(t_i)| \tag{58}$$

$$< \sum_1 |q^m(t_i)| + [2r\alpha_l + (\frac{r}{r+1})^r \frac{1}{r+1} + \epsilon] \sum_{i=\varsigma}^{2^n-1} P(B^m_{t_i,t_{i+1}})$$

$$+ \sum_{i=\varsigma}^{2^n-1} P(N^m_n(t_i, t_{i+1}) > 0,\ N^m_n(t_i - l, t_i) > 0)$$

here $t_\varsigma \in \{t_i\}_{i=0}^{2^n}$ is the rightmost point contained in the first partition element, $[0, \frac{T_m}{K}]$, of the $K$-th order partition. Now applying the relation (31) and Lemma 1.2 to the first term to the right of the inequality (58) for $n$ sufficiently large

$$\sum_1 |q^m(t_i)| \le \sum_1 P(B^m_{t_i,t_{i+1}}) \le \mathbf{E}\,[N^m_{r_1}] + \epsilon \tag{59}$$

where $N^m_{r_k}$ is as defined in Eq. (42). Likewise

$$\sum_{i=\varsigma}^{2^n-1} P(B^m_{t_i,t_{i+1}}) \le \sum_{i=0}^{2^n-1} P(B_{t_i,t_{i+1}}) \le \mathbf{E}\,[N^m(T_m)] + \epsilon \tag{60}$$

which gives via Eq. (58)

$$\lim_{n\to\infty} \sum_{i=0}^{2^n-1} |q^m(t_i)|$$

$$< \lim_{n\to\infty} \mathbf{E}\,[N^m_{r_1}] + [2r\alpha_{m,l} + (\frac{r}{r+1})^r \frac{1}{r+1} + \epsilon]\{\mathbf{E}\,[N^m(T_m)]\} \tag{61}$$

$$+ \lim_{n\to\infty} \sum_{i=\varsigma}^{2^n-1} P(N^m_n(t_i, t_{i+1}) > 0,\ N^m_n(t_i - l, t_i) > 0)$$

Therefore taking the limit as $m, l \to \infty$, $l = o(T_m)$, the first term to the right of (61) goes to zero by Uniform Negligibility, (42), and the finiteness of $\mathbf{E}\,[N^m(T_m)] = \mathbf{E}\,[\widetilde{N}(1)] = \int_0^1 \rho(\tau)d\tau$. The second term converges to a quantity not exceeding $[\frac{1}{r} + \epsilon]\,\mathbf{E}\,[\widetilde{N}(1)]$. However as $m$ becomes unbounded $r$ can be made arbitrarily large and $\epsilon$ can be made arbitrarily small, by Lemma 2.2, thus the second term is negligible. Finally the rarefaction condition, Eq. (38), asserts that the third term vanishes. Hence by Proposition 2.1, for $I = [0, t]$

$$p_m(t) \rightarrow p^*(t) = 1 - \exp(-\int_0^t \rho(\tau)d\tau) \quad , \quad t \in [0,1) \tag{62}$$

∎

Proposition 2.2 asserts that the probability that the normalized upcrossing process $\widetilde{N}^m$ is greater than zero in any interval contained in [0,1) is the same as the corresponding probability for a Poisson counting process $N^*$ in the limit as $m \rightarrow \infty$. To show the stronger result that $\widetilde{N}^m$ actually converges in distribution to a Poisson process we will follow Leadbetter [12] in making use of a theorem in [9]. Using the nomenclature in [9] a point process $N$ is *a-regular* if for every collection of intervals $I$ contained in $\Upsilon_{[0,1]}$, the Borel sets on [0,1], there exists some array $\{I_{mk}\} \subset \Upsilon_{[0,1]}$ of finite covers of $I$ (one for each $m = 1,2, \cdots$) such that

$$\lim_{n \rightarrow \infty} \limsup_{m \rightarrow \infty} \sum_k P(N^m(I_{nk}) \geq a) = 0 \tag{63}$$

We state the following special case of Theorem 4.7 in [9].

*Lemma 2.3*

*Let $\widetilde{N}^m$ be a sequence of point processes and $N^*$ a Poisson process both defined on* [0,1). *Then $\widetilde{N}^m$ converges in distribution to $N^*$ if and only if $\widetilde{N}^m$ is 2-regular and*

$$\lim_{m \rightarrow \infty} P(\widetilde{N}^m(U) = 0) = P(N^*(U) = 0) \tag{64}$$

*for all $U$ of the form*

$$U = \bigcup_{k=1}^{r} \Upsilon_i \quad , \quad \Upsilon_i \subset \Upsilon_{[0,1]}$$

*for $r > 1$, and*

$$\limsup_{m \rightarrow \infty} \mathbf{E}\,[\widetilde{N}^m(I)] \leq \mathbf{E}\,[N^*(I)] < \infty \tag{65}$$

*for $I \subset \Upsilon_{[0,1]}$*

We now proceed to the proof of Theorem 2.1 which at this point only involves showing that $\widetilde{N}^m$ of the theorem satisfies the conditions in Lemma 2.3.

*Proof of Theorem 2.1*

Without loss of generality we assume that the collection of intervals $I$ in the *a-regularity* condition and in (65), and the $\Upsilon$ in (64) are sets of disjoint intervals. For each $m$, $m = 1, \cdots$, define the increasing set of disjoint covers of $I$: $\{J_{mk}\}$, $k = 1, 2, \cdots r_m$, with each $J_{mk}$ of length $l_m / T_m$ (recall $l_m = o(T_m)$). Assume for definiteness that $J_{mk}$ are ordered such that the left endpoints are strictly increasing as $k$ increases. With $N^m$ as in

Theorem 2.1 and $\widetilde{N}^m(\tau) = N^m(\tau T_m)$ we have

$$\sum_{k=1}^{r_m} P(\widetilde{N}^m(J_{mk}) > 1) = \sum_{k=2}^{r_m} P(\widetilde{N}^m(J_{mk}) > 1) + P(\widetilde{N}^m(J_{m1}) > 1) \tag{66}$$

$$\leq \lim_{n\to\infty} \sum_{k=1}^{r_m} \sum_{t_i \in J_{km} T_m} P(N_n^m(t_i^k, t_{i+1}^k) > 0,\ N_n^m(t_i^k - l_m, t_i^k) > 0) + \mathbf{E}\,[\widetilde{N}^m(J_{m1})]$$

where $\{t_i^k\}_{i=1}^{n^k}$ are increasingly dense partitions of $J_{mk}$, for $k = 1, \cdots r_m$ respectively. The first term on the right of the inequality (66) is bounded by the expression in the Rarefaction condition, Eq. (38) while the second term converges to zero by Uniform Negligibility, (42), and the finiteness of $\mathbf{E}\,[\widetilde{N}^m]$. Taking the limit of Eq. (66) as $m \to \infty$ we have that $\widetilde{N}^m$ is 2-*regular* in the sense of (63).

Fix $r > 0$. Because of the absolute continuity of the distributions of $X_m$ the intervals $\Upsilon_i$ in (64) can be taken as having no common boundary points. Therefore by mixing, Lemma 2.1, for any collection of disjoint intervals $\Upsilon_1, \Upsilon_2, \ldots, \Upsilon_r$ contained in [0,1]

$$|P(\bigcap_{i=1}^{r} N^m(T_m \Upsilon_i) > 0) - \prod_{i=1}^{r} P(N^m(T_m \Upsilon_i) > 0)| \to 0\,, \quad as \ m \to \infty \tag{67}$$

where we have adopted the operator notation for $T_m$: $T_m[\sigma,\nu) = [T_m\sigma, T_m\nu)$ for $0 \leq \sigma < \nu < 1$. Eq. (67) and Proposition 2.2 thus imply that

$$P(\bigcap_{i=1}^{r} N^m(T_m \Upsilon_i) = 0) \to \exp(-\sum_{i=1}^{r} \int_{\Upsilon_i} \rho(\tau) d\tau) \tag{68}$$

Since, furthermore, $\widetilde{N}^m$ and $N^*$ have identical intensity (recall (33)) the assumptions stated in Lemma 2.3 are valid for $\widetilde{N}^m$ and Proposition 2.1 establishes Theorem 2.1.

∎

While the asymptotic theorem, Theorem 2.1, is an interesting result, Lemma 2.2 is more useful in applications. Let $N$ be an upcrossing count process, on [0,1], with (incomplete) intensity $\rho$. Lemma 2.2 states a bound on the "approximation error" of the Poisson model, $p^*(t) \triangleq P(N^*(t) > 0)$ and $p(t) \triangleq P(N(t) > 0)$.

$$|\, p^*(t) - p(t) \,| \leq \int_0^1 |\, q(\tau) \,|\, d\tau \exp(\int_0^1 \rho(\tau) d\tau) \tag{69}$$

where the abstract integral has been defined

$$\int_0^1 q(\tau) d\tau \triangleq \lim_{n\to\infty} \sum_{i=1}^{2^n-1} |\, q(t_i) \,| \tag{70}$$

Along with definition (12) for $q$, (70) asserts that as upcrossings become rare over time, $q \to 0$ and there is progressively smaller error involved in the Poisson approximation. Since

$q(\tau) \leq \rho(\tau)$ a uniform decrease in the intensity over time is sufficient for rarefaction. In the next section level crossings will correspond to large errors in an estimation problem involving signals and noises. In that context it can be shown that $\rho$ is monotone decreasing in signal-to-noise-ratio. Thus a practical interpretation of Theorem 2.2 is that a Poisson model for large error is accurate for moderately large SNR and above. Furthermore, it can be shown that, for large $\rho$, $p^*(t) \geq p(t)$. This establishes $p^*$ as an upper bound for $p$ under large or small SNR conditions.

## IV. APPLICATION TO PASSIVE ARRAYS

The performance of the correlator estimate of time delay in a two sensor passive array has received much attention in the past decade [2,20,10]. As is typical in non-linear estimation problems, exact expressions for the variance of any estimator are difficult to derive except under restrictive small error regimes [17]. In this section we will develop a global variance approximation which is directly motivated by a level crossing interpretation for large errors. In this context the asymptotic result presented in the last section has an interesting interpretation. For low probability of large error, the level crossings form a point process over the a priori region which have nearly Poisson statistics. While for large probability of error the Poisson model is conservative, i.e. $P(N>0)$ is larger for a Poisson $N$ than for the actual level crossing process $N$. This observation suggests building a conservative global variance estimate, via Poisson modeling, to complement the lower bounds such as the Ziv-Zakai and Cramer-Rao bounds.

Our observation model for the outputs of a two-sensor passive array is as follows. The outputs of two sensors, $x_1(t)$ and $x_2(t)$, are observed over a finite interval of time $t \in [0, T]$

$$\begin{aligned} x_1(t) &= s(t) + n_1(t) \\ x_2(t) &= s(t-D) + n_2(t) \end{aligned} \tag{71}$$

The signal components, $s(t)$ and a delayed version $s(t-D)$, and the noises, $n_1(t)$ and $n_2(t)$, are zero mean, uncorrelated, stationary Gaussian random processes. The delay $D$ is restricted to an *a priori* region of possible delay $[-D_M, D_M]$. The signal auto-correlation, $R_{ss}(\tau) = \mathrm{E}\,[s(t)s(t+\tau)]$, is assumed to be essentially zero for $|\tau| > T_c$, where $T_c = k/B$ is the correlation time, $k$ is an integer and $B$ is the baseband bandwidth of the signal. As in most cases of interest, we assume that the uncertainty region $[-D_M, D_M]$ is large enough to make the time-bandwidth product $BD_M >> 1$.

For flat signal and noise spectral densities, the correlation estimate of time delay, $\hat{D}$, is the location in time, within $[-D_M, D_M]$, at which the global maximum of the sample cross-correlation function occurs.

$$\hat{D} \triangleq \underset{\tau \in [-D_M, D_M]}{\mathrm{argmax}} \hat{R}_{12}(\tau) \tag{72}$$

$\hat{R}_{12}(\tau)$ is the sample cross-correlation function.

$$\hat{R}_{12}(\tau) \triangleq \frac{1}{T} \int_0^T x_1(t) x_2(t+\tau) dt \tag{73}$$

The correlator estimate is known to be asymptotically equivalent to the maximum likelihood estimate for $D$ as $BT \rightarrow \infty$[10]. In [15] a simple small error approximation, the Cramer-Rao-Lower-Bound: $\sigma^2_{CRLB}$, was derived for the variance of $\hat{D}$. The CRLB is an accurate approximation to the true variance when $|\hat{D}-D| < \delta$ with high probability, where $\delta$ is a small constant dependent upon the signal and noise spectra. An exact expression for the global variance results from direct application of the "law of total probability" to $var\{\hat{D}\}$:

$$var\{\hat{D}\} = \sigma^2_{loc}(1 - P_e) + \sigma^2 P_e \tag{74}$$

In (74) $\sigma^2_{loc} = \min\{\sigma^2_{CRLB}, \frac{\delta^2}{3}]\}$ and $\sigma^2 = \mathbf{E}\{(\hat{D} - D)^2 \mid (\hat{D} - D) \notin [-\delta,\delta]\}$ are the conditional expectations of the squared error given small (local) error and large error respectively, and $P_e = P(\hat{D} - D \notin [-\delta,\delta])$ is the probability of large error. The rest of this section deals with suitable approximations to the large error probability and the squared error $\sigma^2 Pe$.

The occurrence of a peak equal or greater in magnitude than the local maximum of the correlator within $[-\delta,\delta]$ is called a peak ambiguity, since it confounds the estimators search for the location of the local max occurring near the true delay $D$. A useful interpretation is that each peak ambiguity gives rise to a candidate for $\hat{D}$. In the exact model the candidates are the locations over the a priori interval where the peak ambiguities occur. From these candidates a single member is then selected for $\hat{D}$, the one which corresponds to the largest of the ambiguous peaks. We use the above interpretation to develop a different set of candidates, each of which corresponds to a level crossing location associated with each peak ambiguity. $\hat{D}$ is then modeled as equally likely to take on the identities of any of the candidates. With little loss of generality it will be assumed that $D=0$ in the sequel [5].

Define the random level $m_\delta \triangleq \max_{u \in [-\delta,\delta]} \hat{R}_{12}(u)$ and the "ambiguity process" $\Delta R(\tau) \triangleq \hat{R}_{12}(\tau) - m_\delta$. $m_\delta$ is the magnitude of the desired local peak of the sample cross-correlation, while $\Delta R(\tau)$ must be negative over $[-D_M, D_M] - [-\delta,\delta]$ for no large error to occur. Next define the level crossing count process $N \triangleq \{N(\tau):\tau \in [-D_M, D_M]\}$ associated with the ambiguity process

$$N(\tau) \triangleq \begin{cases} N_u(-D_M,\tau) & \tau \in [-D_M,-\delta) \\ N_u(-D_M,-\delta) & \tau \in [-\delta,\delta] \\ N_u(-D_M,-\delta) + N_d(\delta,\tau) & ,\ \tau \in (\delta, D_M] \end{cases} \tag{75}$$

where $N_u(t_1,t_2)$ is the number of ***up-crossings*** of zero by $\Delta R(\tau)$ over $\tau \in [t_1,t_2) \subset [-D_M,-\delta)$, and $N_d(t_1,t_2)$ is the number of ***down-crossings*** of zero by $\Delta R(\tau)$ over $\tau \in [t_1,t_2] \subset (\delta, D_M]$. The process $N$ is merely the running sum, over time, of the total number of up-crossings to the left of the true delay plus down-crossings to the right of the true delay.

With the above definitions, no large error occurs, i.e. $\hat{D} \in [-\delta,\delta]$, if and only if 1). $\Delta R(-D_M) < 0$ and $\Delta R(D_M) < 0$ and 2). $N(D_M) = 0$. Hence the probability of large error

$$P_e = 1 - P(N(D_M) = 0, \Delta R(-D_M) < 0, \Delta R(D_M) < 0) \tag{76}$$

Define the left continuous probability distribution function $F(x) \triangleq P(\Delta R(-D_M) < x)$. The probability of large error (76) can be shown to have the form [6]

$$P_e = 1 - \exp\{-\int_{-D_M}^{D_M} \lambda_c(\tau) d\tau\} \; F^2(0) \tag{77}$$

Where $\lambda_c$ is the conditional (incomplete) intensity of $N$, $\lambda_c(\tau) d\tau \triangleq P(dN(\tau) > 0 \mid N(\tau)=0, \Delta R(-D_M) < 0)$ and $dN(\tau)$ is the infinitesimal increment in time of $N$ at time $\tau$.

Equation (77) is an exact representation of the probability of large error in terms of the level crossings $N$. However, while the presence of level crossings is (conditionally) equivalent to the presence of ambiguity, the level crossings alone do not uniquely specify the location of the global maximum. Hence one cannot expect the level crossings to be sufficient, by themselves, to give an exact expression for variance. As an approximation, we will use the following conditionally uniform model for the location of the global maximum given a particular sequence of level crossings. Let $\{\omega_1, \ldots, \omega_n\}$ be the ordered set of points in $[-D_M, D_M]$ where $N(\tau)$ increases. Conditioned on the occurrence of a large error, we will model $\hat{D}$ as follows: $\hat{D}$ takes values in the set $\{\omega_1, \ldots, \omega_n\}$ with equal probability if $N(D_M) = n$ and both $\Delta R(-D_M)$, $\Delta R(D_M) < 0$; while $\hat{D}$ takes values in one of the sets: $\{\omega_1, \ldots, \omega_n, \pm D_M\}$, $\{\omega_1, \ldots, \omega_n, -D_M, D_M\}$, with equal probability if $N(D_M) = n$ and either: $\Delta R(-D_M) \geq 0$, $\Delta R(D_M) \geq 0$; or both, respectively.

Assume for simplicity that $N_d(\delta, D_M) = 0$ while $Nu(-D_m, -\delta) > 0$. The conclusions drawn for this case apply to the more general situation with no additional conceptual difficulty. Define $\{a_1, \ldots, a_N\} \triangleq \{\underset{[w_1,w_2)}{argmax}\Delta R(\tau), \ldots, \underset{[w_N,-\delta)}{argmax}\Delta R(\tau)\}$ the ordered set of peak ambiguity locations. Since $|a_i| \leq |w_i|$, $i = 1, \ldots, N$, and the largest ambiguities tend to cluster in the vicinity of the high amplitude sidelobes of $R_{ss}$, the signal autocorrelation, occurring close to the true delay, the proposed model entails, at worst, a pessimistic estimate of the mean squared error of $\hat{D}$.

Under the conservative model described in the preceding paragraphs, the following inequality can be derived [6].

$$var\{\hat{D}\} \geq \sigma_{loc}^2 \exp\{-\int_{-D_M}^{D_M} \lambda_c(\tau)d\tau\} + \int_{-D_M}^{D_M} \tau^2 \rho(\tau)\, g(\tau)d\tau + D_M^2\alpha \qquad (78)$$

In (78) $\rho$ is the unconditional (incomplete) intensity of $N$, as defined in Section II, Eq. (9), and $\alpha$ is a small quantity given in [5] which is not significant in this discussion. The function $g(\tau)$ is defined as

$$g(\tau) \triangleq \sum_{n=1}^{\infty} \frac{1}{n} \sum_{k=1}^{n} a_{k-1,n-k}(\tau) \qquad (79)$$

and we have defined the bidirectional Palm measure

$$a_{k-1,n-k} \triangleq \lim_{h \to 0} \frac{1}{h}\, P(N(\tau) = k-1,\, N(D_M) = n \mid dN(\tau,h) > 0) \qquad (80)$$

The bidirectional Palm measure corresponds to the probability that, given the occurrence of a (crossing) point at $\tau$, this point is the $k$-$th$ occurrence in a sequence of $n$ points occurring over $[-D_M, D_M]$ (see [3] for a discussion of Palm measures).

The expression (78) consists of three factors. The first factor is the small error contribution to the global variance. The second term is the contribution of peak ambiguities which generate level crossings in $(-D_m, D_m)$, and the third term is the contribution of any peak ambiguity which does not generate a level crossing (i.e. corresponding to our conservative assignment $\hat{D} = \pm D_M$).

The asymptotic results cited in the previous sections suggest the feasibility of applying a Poisson approximation to the level crossing process $N$ as these crossings become increasingly rare, i.e. for small $\rho$ (Recall discussion at the end of Section III). Here we give more quantitative results concerning the actual error committed by the approximation. Let $N^*$ be an inhomogeneous Poisson process with intensity $\rho(\tau)$ and define the signed difference $\Delta \triangleq P(N^*(\sigma,u) > 0) - P(N(\sigma,u) > 0)$. While sharp bounds on the deviation of $\Delta$ from zero can be derived for $\Delta R$ a nonstationary Gaussian process, using Theorem 1.1, we will concentrate on the following non-parametric bounds derived in [6].

$$\max\left\{-\frac{\mathbf{E}^2\{N\}}{1+\mathbf{E}\{N\}},\, -e^{-\mathbf{E}\{N\}}\right\} \leq \Delta \leq \mathbf{E}\{\frac{1}{N+1}\} - e^{-\mathbf{E}\{N\}} \qquad (81)$$

Where, for compactness, $N$ is shorthand for $N(\sigma,u)$. Note that all terms in the left and right hand inequalities of (81) depend on only the first moment of $N(\sigma,u)$, except for $\mathbf{E}\{\frac{1}{N+1}\}$. This latter term can be upper bounded, however

$$\mathbf{E}\{\frac{1}{N+1}\} \leq \frac{G\{N\}}{1+G\{N\}} \tag{82}$$

where

$$G\{N\} \triangleq \frac{var\{N\}}{\mathbf{E}^2\{N\}} + \frac{1}{\mathbf{E}\{N\}} \tag{83}$$

The mean and variance of $N(\sigma,u)$ can be computed from the second and fourth order distributions of $X$ if Leadbetters conditions are satisfied [11]. Thus if the variance of $N(\sigma,u)$ increases only as rapidly as $o(\mathbf{E}^2\{N(\sigma,u)\})$ the bounds (81) and (82) guarantee that the error incurred in the Poisson approximation is near zero whenever $\exp\{-\mathbf{E}\{N(\sigma,u)\}\}$ approaches either 1 or 0. In any case, for these extremal conditions the left hand inequality in (81) implies $\Delta$ is lower bounded by a small magnitude negative quantity, i.e. the Poisson model conserves the inequality (78) to a good approximation.

The application of the Poisson model to the level crossing process $N$ gives the following simple relations for the probability of large error and the bound on the variance (78)

$$P_e = 1 - e^{-\mathbf{E}_c\{N(D_M)\}F^2(0)} \tag{84}$$

$$var\{\hat{D}\} \leq \sigma_{loc}(1-P_e) + \int_{-D_M}^{D_M} \tau^2 \hat{\rho}(\tau)d\tau \; (1 - e^{-\mathbf{E}\{N(D_M)\}}) + D_M^2\alpha \tag{85}$$

In (85) we have defined the normalized intensity

$$\hat{\rho}(\tau) \triangleq \rho(\tau)/\int_{-D_M}^{D_M} \rho(u)du \tag{86}$$

and $\alpha$ is given by

$$\alpha \triangleq 2\left[\frac{1 - e^{-\mathbf{E}\{N(D_M)\}}}{\mathbf{E}\{N(D_M)\}} - \frac{1 - e^{-\mathbf{E}_c\{N(D_M)\}}}{\mathbf{E}_c\{N(D_M)\}}(1-F^2(0))\right]$$

Here

$$\mathbf{E}_c\{N(D_M)\} = \int_{-D_M}^{D_M} \rho_c(\tau)d\tau \;, \quad \mathbf{E}\{N(D_M)\} = \int_{-D_M}^{D_M} \rho(\tau)d\tau \tag{87}$$

are the conditional mean of $N(D_m)$ given $\Delta R(-D_m) > 0$ and the unconditional mean of $N(D_M)$ respectively.

The Poisson approximation (85) indicates that, as the intensity of peak ambiguities, $\rho$, increases, one must discount the small error variance, $\sigma_{loc}^2$ by $P_e$, adding an increasingly large quantity: the mean-square deviation of the locations of peak ambiguities overtime. In the following section we will explicitly calculate the intensities in (87) under a Gaussian assumption, and analyze the resulting form of the Poisson variance approximation for simple bandpass signals.

V. NUMERICAL COMPARISONS

The intensity functions $\rho_c$ and $\rho$ in (87) can be derived under the following assumptions: a). $\hat{R}_{12}$ is a Gaussian random process with non-stationary mean and differentiable covariance function; and b). $\max_{u \in [-\delta,\delta]} \hat{R}_{12}(u) = \hat{R}_{12}(0)$. The Gaussian model is reasonable for large BT [5]. Since the exceedance of $\hat{R}_{12}(0)$ by $\hat{R}_{12}(\tau)$ for some $\tau \in [-D_M, D_M]-[-\delta,\delta]$ does not necessarily imply a peak ambiguity, assumption b). is pessimistic at worst.

Using the assumptions a). and b). the results are

$$\rho_c(\tau) = K_1 \int_0^\infty y \; \Phi(a_0 y + a_1) \; \phi(y + a_2) dy \tag{88}$$

$$\rho(\tau) = K_2 \; \phi(a_3)[\phi(a_4) + a_4 \Phi(a_4)]$$

Here $K_1, K_2, a_0, \ldots, a_4$ are functions of $\tau$ given [6]. The functions $\Phi$ and $\phi$ are the standard Gaussian distribution and density functions respectively.

In [6] a simple explicit forms for (88), (84) and (85) was derived for flat lowpass signal and noise spectra. For these simple bandpass spectra the small error region over which the CRLB is accurate $[-\delta,\delta]$ is given by $\delta = 1/4f_o$. Here we only discuss numerical results for flat bandpass spectra. In Fig. 1 the intensity surface, is displayed for a bandpass signal at center frequency $f_o = 500Hz$, with bandwidth $B = 200Hz$, and $T = 8.0secs$. Here the time window extends from the first zero crossing of the auto-correlation function of the signal at $\delta = 1/4f_o$, to approximately the fifth sidelobe away from the origin. In Fig. 1 the location of the global maximum of the autocorrelation is just beyond the rightmost point on the $t$ axis. A distinctive feature of Fig. 1 is the SNR difference between the point, $SNR_1$ where a rapid rise in the intensity of ambiguity first begins, i.e. in the region of the first sidelobe, and the point, $SNR_2$ where a uniform increase of the ambiguity, over time, is in evidence. This implies the existence of at least three separate SNR thresholds which is consistent with studies of the Ziv-Zakai-Lower-Bound (ZZLB) for this problem [20].

We numerically evaluated the integrals in (88) and (85) for a flat bandpass signal with center frequency to bandwidth ratio $f_o/B = 10$, and $BD_M = 25$. The results are plotted in Figs. 2 and 3, along with plots of the CRLB and ZZLB, for $BT = 200$ and $BT = 80$ respectively. The Poisson approximation behaves similarly to the ZZLB in Fig. 2, both indicating the presence of three distinct SNR thresholds (e.g. $SNR_{t1}$, $SNR_{t2}$ and $SNR_{t3}$ in Fig. 2) of performance. For $SNR < SNR_{t1}$ the Poisson approximation becomes a much better predictor of variance than the CRLB. $[SNR_{t2}, SNR_{t1}]$ is a region where, with high probability, large errors are concentrated in the interval $\hat{D} \in [-T_c, T_c]$, the small error region for the envelope of the bandpass signal. When $SNR < SNR_{t2}$ the error approaches that of a uniform random variable over $[-D_M, D_M]$: the estimate $\hat{D}$ is useless. For $BT=80$, in Fig. 3 the Poisson approximation has moved away from the ZZLB relative to the case of $BT=200$. Indeed it appears to hit an asymptote with increasing SNR, i.e. the correlator commits large errors even as the SNR approaches infinity. This behavior of the Poisson approximation corroborates the reported sub-optimality of the correlator estimate for small BT [7].

Finally the results of a simulation of correlator performance for a bandpass signal spectrum appears in Fig. 4. The relevant parameters are: $f_0/B = 2.5$, $BT = 50$ and $BD_M = 8$ and the vertical dimension of the "Φ" characters indicate approximate 95% confidence interval for the actual variance (obtained by simulation). Plotted for comparison are the CRLB, ZZLB and Poisson Approximation. The combination of the overly optimistic ZZLB and the overly pessimistic Poisson approximation jointly specify an admissible region of estimator variance. However, on the average, below a SNR of 5dB the Poisson approximation is significantly closer to the true variance than the ZZLB. Note in particular that at a SNR of -8dB the Poisson approximation is within the 95% confidence interval while the ZZLB is more than 5dB below this interval.

## VI. CONCLUSION

Two results were derived in the context of level crossing probabilities. First, a representation of the probability of getting one or more upcrossings in an interval by a general random process was presented. This representation in effect isolates the portion of the upcrossing probability due to the intensity function of the upcrossings, from a correction term, which characterizes the deviation of the upcrossing probability from an associated inhomogeneous Poisson probability. The correction term depends on the degree to which the upcrossings can be modeled as an independent increment process. By identifying conditions which asymptotically force the correction term to zero a second result was made possible: that a certain time normalized version of the upcrossing process can be made to converge in distribution to the inhomogeneous Poisson law.

Future investigations of the of the correction term, $Q(t)$, associated with the probability representation of Theorem 1.1, should lead to useful expressions for the approximation error incurred by using such simple first moment approximations. For the asymptotic result, Thm. 2.1, the asymptotic conditions rarefaction and mixing play an important role. In particular, rarefaction could be replaced by conditions involving probability statements about the maximum process over the interval $I$, $\max_{\tau \in I} X(\tau)$, analogous to [12]. For specific probability models of the random process $X(t)$ of interest, e.g. Gauss-Markov or Rayleigh as in [12], one would expect the replacement condition to be more easily verified, than the conditions used here.

An application of the Poisson model to a problem in underwater acoustics, time delay estimation, yielded an approximation to the global variance of the estimate. Numerical results indicate the fidelity and conservativeness of this performance approximation relative to the Ziv-Zakai lower bound for bandpass spectra. Yet to be investigated is the feasibility of Poisson approximations to large error in multiparameter estimation problems In these situations the maximum likelihood procedure involves a search for a global maximum over an ambiguity surface. Thus the concept of level crossing becomes more difficult due to the lack of inherent directionality over the parameter space (points in the space are not well ordered). While this would not preclude the application of a Poisson spatial model for the locations of peak ambiguity, bounds on the approximation error, analogous to the one dimensional case, are not as simple to derive.

ACKNOWLEDGEMENT

This research was supported in part by the Office of Naval Research under contract N00014-81-K-0146 and by a 1985 Rackham Research Grant from the graduate school of the University of Michigan, Ann Arbor.

## REFERENCES

1. I.F. Blake and W.C. Lindsay, "Level-crossing problems for random processes," *IEEE Trans. on Inform. Theory,* Vol. IT-19, pp. 295-315, 1973.
2. S.K. Chow and P.M. Schultheiss, "Delay estimation using narrowband processes," *IEEE Trans. Acoust., Speech, Signal Proc.*, Vol. ASSP-29, No. 3, pp. 478-484, 1981.
3. D.R. Cox and V. Isham, *Point Processes* Chapman and Hall, N.Y., 1980.
4. H. Cramer, *Mathematical Methods of Statistics,* Princeton University Press, Princeton, NJ, 1951.
5. A. Hero, *Topics in time delay estimation,* Ph.D dissertation, Princeton University, Princeton, N.J. 08544, 1984.
6. A. Hero and S.C. Schwartz, "Poisson models and global variance for passive time delay estimation," submitted for publication to IEEE Trans. on Inform. Theory, Sept. 1985.
7. J.P. Ianniello, E. Weinstein, and A. Weiss, "Comparison of the Ziv-Zakai lower bound on time delay estimation with correlator performance," *ICASSP-83 Proceedings,* Vol. 2 , pp. 875-878.
8. M. Kac and D. Slepian, "Large excursions of Gaussian processes," *Ann. of Math. Statist.* Vol. 18, pp. 383-397, 1947b.
9. O. Kallenberg, Random Measures, Berlin, Akademie-Verlag, 1975.
10. C.H. Knapp and G.C. Carter, "The generalized correlation method for estimation of time delay," *IEEE Trans. Acoust., Speech, Sig. Proc.*, Vol. ASSP-24, No. 4, pp. 320-327, 1976.
11. M.R. Leadbetter, "On Crossings of Levels and Curves by a Wide Class of Stochastic Processes", *Ann. of Math. Statist.* Vol 37, pp. 260-267, 1966.
12. M.R. Leadbetter, "Extreme value theory for continuous parameter stationary processes", *Zeitschrift fur Wahrscheinlichkeitstheorie und Verwandte Gebiete,* Vol. 60, pp. 1-20, 1982.
13. M.R. Leadbetter, *"Point processes generated by level crossings,"* P. A. W. Lewis, ed., Stochastic Point Processes, New York, Wiley, 1972.
14. M.R. Leadbetter, G. Lindgren and H. Rootzen, *Extremes and related properties of random sequences and processes,* Springer-Verlag, New York, 1982.
15. V.H. MacDonald and P.M. Schultheiss, "Optimum passive bearing estimation in a spatially incoherent noise environment", *Journal of the Acoustical Society of America,* Vol. 46, No. 1, Part 1, pp. 37-43, 1969.
16. M. Rosenblatt, Random Processes, N.Y., Oxford University Press, 1962.

17. L.P. Seidman, "An upper bound on average estimation error in nonlinear systems," *IEEE Trans. Inform. Theory,* Nov 71, pp 655-665.

18. M. Vidyasagar, Nonlinear Systems Analysis, Englewood Cliffs, Prentice-Hall, 1978.

19. V.A. Volkonskii and Y. A. Rozanov, "Some limit theorems for random functions. II," *Theory of Prob. and its App. (English Translation),* Vol. 6, No. 2, pp. 186-198, 1961.

20. E. Weinstein and A.J. Weiss, "Fundamental limitations in passive time delay estimation - part II: wide-band systems" *IEEE Trans. Acoust., Speech, Signal Proc.,* Vol. ASSP-32, No. 5, pp. 1064-1077, 1984.

21. N.D. Ylvisaker, "The expected number of zeros of a stationary Gaussian Process", *Ann. of Math. Statist.,* Vol 36 pp 1043-1046, 1965.

22. P. Billingsly, *Probability and Measure,* Wiley, New York, 1979.

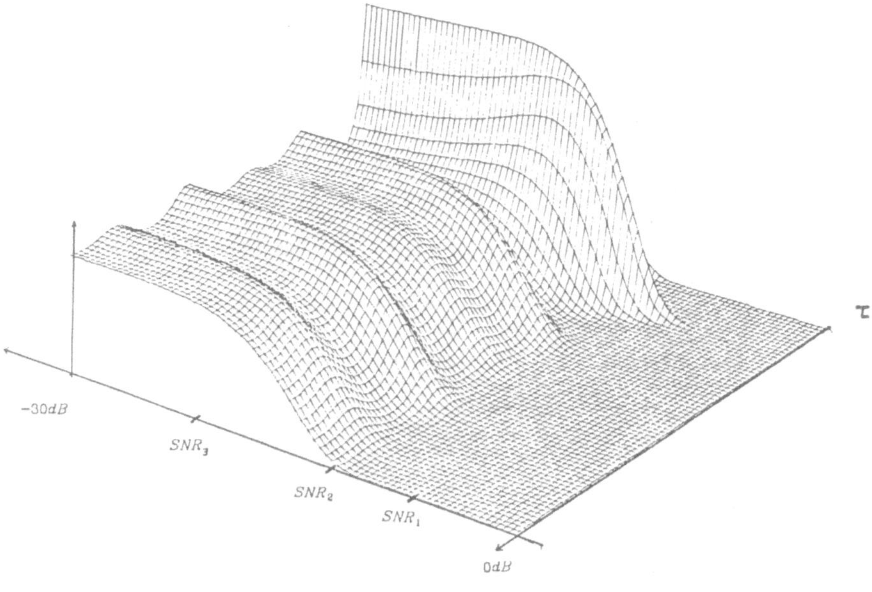


**FIG. 1**
Intensity surface, λ, for bandpass signal over time and SNR.

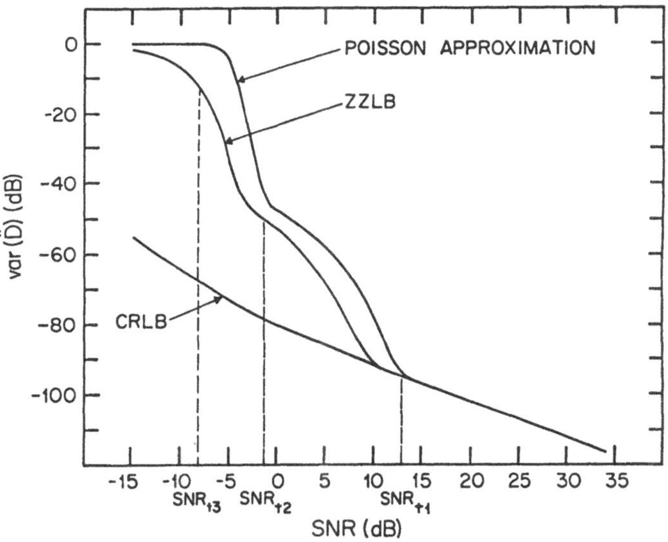


FIG. 2
Comparison of Poisson approximation with ZZLB and CRLB for bandpass signal spectrum. $f_o/B = 10$, $BD_m = 25$ and $BT = 200$. Variance, $var(\hat{D})$, normalized with respect to standard uniform distribution.

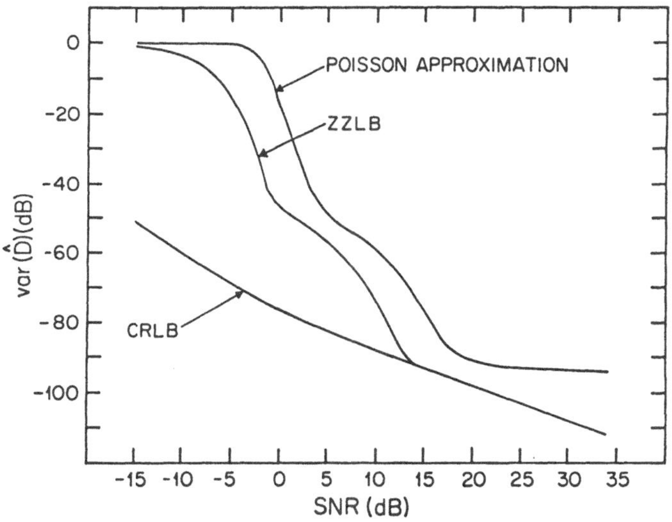


FIG. 3
Comparison of Poisson approximation with ZZLB and CRLB for bandpass signal spectrum. $f_o/B = 10$, $BD_m = 25$ and $BT = 80$.

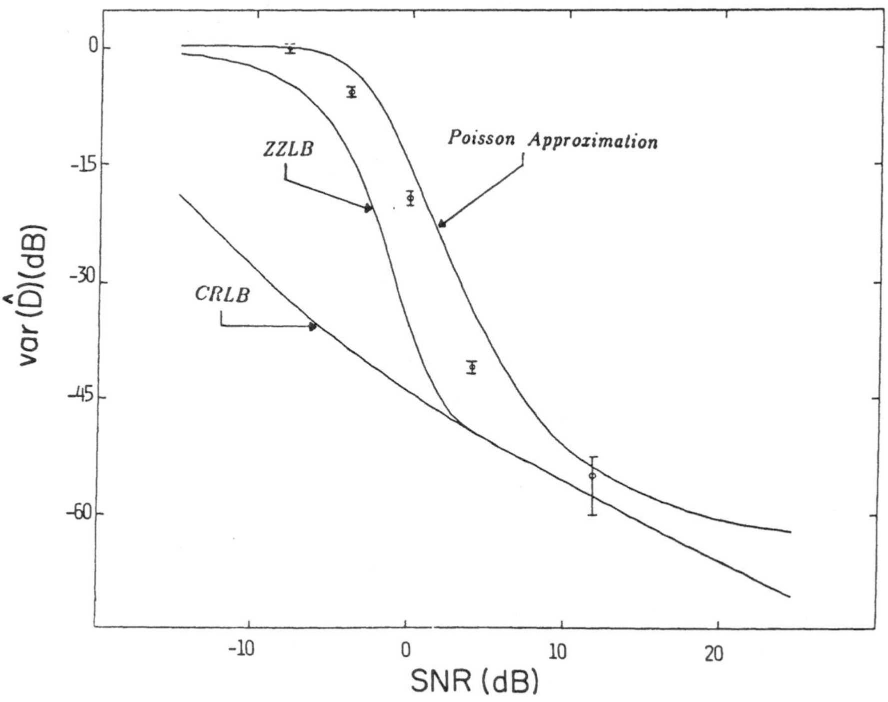


FIG. 4
Comparison of Poisson approximation with ZZLB and results of simulation, Φ, for $f_o/B = 2.5$, $BD_m = 8$ and $BT = 50$.

CHAPTER 5

# NONLINEAR DATA OBSERVABILITY AND NONGAUSSIAN INFORMATION STRUCTURES

R.R. Mohler and C.S. Hwang

## 1. INTRODUCTION

Here data observability refers to the ability to reconstruct a process state x(t) from measured data y(t), where $t \in [t_0, t_1]$. The receiver is modeled by

$$y(t) = h(x) \ , \tag{1}$$

where x(t), an unknown process state (such as acoustic source location and motion), is given by

$$\frac{dx}{dt} = f(x) \ , \tag{2}$$

$x(t) \in R^n$, $y(t) \in R^m$, and $f(\cdot)$, $h(\cdot)$ are appropriate n and m nonlinear functionals, respectively. Equation (2) assumes at least a certain intelligent guess as to the structure of the data source such as a submerged moving body.

For (1) and (2) linear in x(t), well-developed test criteria can be used to determine observability [1,2]. But for the general nonlinear process state observability is, in general, much more complicated. In a geometric sense, a functional relationship between measurement space and state space is not generally one-to-one such that an inverse function between these two spaces is unique.

If the system involves noise processes in (1) and (2), then the situation becomes even more complicated. In this case, a "yes" or "no" answer generally cannot be given but a degree of "observability" may be given in terms of observed information about the state.

For stochastic observability, an information theoretical point of view will be studied, i.e., to find the quantity of information contained in the observation process. More precisely, "mutual information" between the a priori unobserved dynamic process and the a posteriori observed process will be used as a criterion of the data or system observability.

In Section 2, deterministic nonlinear observability is discussed. A global version of the inverse function theorem proves a base for two observability conditions: connectedness and univalence conditions between the state space and the measurement space. Several signal processing examples, including: i) a BOT (bearing-only target) which is described by a mixed-coordinate system and ii) a linear-array SONAR-tracking problem with delay or Doppler measurement, are studied.

In Section 3, stochastic nonlinear "observability" is studied in relative terms by mutual information. Since this requires a computation of entropy difference between a priori and a posteriori processes estimation theory is useful.

Since an exact nonlinear estimation requires infinite dimensions, a proper finite-dimensional approximation is used in the practical implementation. Then mutual information may be expressed in a logarithm of a determinant ratio of the covariance of the conditional and unconditional processes as far as their mean and covariance expressions are available. Since information in the Shannon sense is used, the Gaussian case provides an upper bound in the information content which is similar to the familiar Cramer-Rao bound with the Fisher information matrix.

Finally, the effect of deterministic observability to stochastic "observability" is discussed. Roughly speaking, deterministically unobservable or weakly observable states show very slow information growth compared to that of observable states.

Application of the above theoretical approach to practical problems is simulated in Section 4. Information structures of individual states and complete state for the underwater environment are analyzed for the BOT and linearly deployed SONAR array.

For the BOT, three different coordinate systems including rectangular, modified-polar and mixed coordinate (mixed with rectangular and polar components) are compared for maneuvering and nonmaneuvering cases.

In the linear-array SONAR, information structural variation depending on the number of sensors and measurement policy is analyzed, and is compared for up to three-sensor deployment.

## 2. DETERMINISTIC NONLINEAR OBSERVABILITY

Various authors have studied nonlinear observability in different ways. Extension of linear observability criteria to the nonlinear case can be found for control processes in [3] and [4]. Here, an observability matrix rank condition [5,6,7] or Taylor series expansion [8] is used. Since observability really involves an inverse function, a well-known inverse function theorem from analysis is used here. In this approach, the Jacobian matrix of the observation related function plays a central role. From this view [9,10,11,12] may be considered in the same category.

### 2.1 A Modified Form of the Global Inverse Function Theorem

In analysis, an inverse function theorem is widely used to get the local inverse of the function by providing a nonzero determinant of the Jacobian J.

Consider an n real-valued continuous function, $F: x \to Y$, $x \in R^n$, $Y \in R^n$ such that

$$F(x) = Y , \tag{3}$$

where F(x) is a $C^1$ map of $R^n$ onto itself. The global inverse function theorem says that the necessary and sufficient conditions for F(x) to be a $C^1$ - diffeomorphism (i.e., an inverse $F^{-1}$ exists and is also differentiable) are given as follows [13]:

1) $\det J\, F(x) \neq 0$ , and (4)

2) $\lim\limits_{\|x\| \to \infty} \|F(x)\| = \infty$ for all x , (5)

where $\|\cdot\|$ is an Euclidean norm.

But the above conditions only guarantee the existence of an inverse function of (3). To be unique for all x and have flexibility in the application, the theorem can be modified. First, consider the following:

<u>Definition</u>

Any individual function of (3), $F_i(x)$ , $i = 1,2,\dots,n$ is called an "absolutely independent function" if it consists of only one component of $x \in R^n$.

Remarks

i) For special cases with F including one absolutely independent function, then

$$\det J F^{-}(x) \neq 0 \text{ for all } x \ ,$$

where $F^{-}$ consists of n-1 functions found by deleting any absolutely independent function from F. And the result can be further generated for more absolutely independent functions. Since the n-dimensional det JF( ) ≠ 0 always includes n-1 dimensional case, the weakened condition will be used only for the special case whenever it applies.

ii) Further restriction of the so called "finite-covering condition" of Palais theorem [13] to a one-covering condition of the current theorem is necessary for F globally one-to-one.

iii) Neither of the conditions alone is enough since the nonzero Jacobian condition alone lacks globality of the inverse G of F, and the one-covering condition alone lacks independence of F.

iv) Since the nonzero Jacobian condition guarantees the existence of only a local inverse, i.e., provides "connectedness" of every component of x to Y, it will be called the connectedness condition in observability analysis. The one-covering condition, on the other hand, provides the uniqueness of this connection. So, this is termed a univalence condition for the observability problem.

## 2.2 Nonlinear Observability

For nonlinear deterministic processes, (1) and (2), it is assumed that f(•) satisfies the required conditions to guarantee the existence and uniqueness of the solution x(t), and y(t) is assumed differentiable up to (n-1)th order in t. Define, then, system observability as follows:

Definition

The process (1), (2) is <u>observable at $t_0$</u> if knowledge of the output measurement y(t), $t \in [t_0, t_1]$ is sufficient to determine $x(t_0)$ uniquely for finite $t_1$. If every state $x(t) \in R^n$ is observable on the time interval considered, then the state is <u>completely observable</u>.

By differentiation of (2) with respect to t and substitution of (1) and with appropriate replacement of lower order derivatives to the higher order successively (with suppression of t in variables for convenience)

$$\begin{aligned} y &= h(x\ t), \\ y' &= h_1(y,\ x,\ t) \\ y'' &= h_2(y,\ y',x), \\ &\cdot \\ &\cdot \\ &\cdot \\ y^{(n-1)} &= h_{n-1}\ (y,y',\ldots,y^{(n-2)},\ . \end{aligned} \tag{6}$$

Denote the vector $Y \varepsilon R^{mn}$ by

$$Y = (y,y',\ldots,y^{(n-1)}) \tag{7}$$

and the set $Y^-$, if needed, by

$$Y^- = \{y,\ y',\ldots,\ y^{(n-2)}\} \tag{8}$$

then the vector notation of (8) becomes

$$Y = H(Y^-,\ x)\ . \tag{9}$$

By the successive replacement of lower-order derivatives to the higher orders, the functional dependency between the individual functional elements h, $h_1$, $h_2$,..., vanishes since this procedure is exactly the same as the successive elimination of unknown variables in solving (6) for x. So, the maximum independency between functional elements is obtained. From (9) and the above theorem, it is readily seen that observability of the system may be determined by the following result.

<u>Observability Result</u>

System (1), (2) is observable (in the strict sense) if (9) satisfies the following conditions for all t, $t \varepsilon [t_0,\ t_1]$:

1) Connectedness condition

$$\det J\ H_n(\cdot) \neq 0\ , \tag{10}$$

where $JH_n = \frac{\partial H_n}{\partial x}$ ; $H_n(\cdot)$ is a subset of $H(\cdot)$ consisting arbitrarily of n of its functions.

2) Univalence condition

For $H_n(\cdot)$, every $x_i$, i=1, 2,..., n, can be uniquely expressed in terms of only Y in (9).

Depending on the satisfaction of the conditions, define and categorize system observability as follows:

1. Observable in the strict sense.
   Both connectedness and univalence conditions are satisfied for any one or more combinations $H_n(\cdot)$ out of mn functions which comprise elements of $H(\cdot)$.
2. Observable in the wide sense.
   Only the connectedness condition is satisfied, i.e., multiple covering appears in some element of x for some time t.
3. Unobservable.
   One or more states of x cannot be expressed in terms of Y. In this case, those states are unconnected to Y and thus to $y_t$.

The method is readily applicable to any linear or nonlinear time-varying as well as invariant system. Several examples follow.

Ex.1, [8], [11]

$$\begin{aligned} \dot{x}_1 &= x_2 x_3 , \\ \dot{x}_2 &= -x_1 x_3 , \\ \dot{x}_3 &= 0 , \end{aligned} \tag{11}$$

$$y = x_1 .$$

Then,

$$\begin{aligned} y' &= x_2 x_3 , \\ y' &= -x_1 x_3^2 = -y x_3^2 , \end{aligned} \tag{12}$$

and

$$J = \begin{bmatrix} 1 & 0 & 0 \\ 0 & x_3 & x_2 \\ 0 & 0 & -2yx_3 \end{bmatrix} .$$

So, det J = $-2yx_3^2 = -2x_1x_3^2 \neq 0$ implies that the initial state of the form $\{x_{10} \neq 0,\ x_{30} \neq 0\}$ satisfies connectedness. But from (12),

$$x_1 = y \ ,$$

$$x_2 = \pm\, y'/\sqrt{\frac{y''}{-y}} \ , \tag{13}$$

$$x_3 = \pm \sqrt{\frac{y''}{-y}} \ .$$

Hence, $x_2$, $x_3$ have multiple covers. So, univalence is not satisfied. The system is observable only in the wide sense if $\{x_{10} \neq 0,\ x_{30} \neq 0\}$ .

Example 2, [14]

Consider an object or target (T) and observer (O) configuration as in Figure 1. When T and/or O move with velocity components $V_{Tx}$, $V_{Ty}$, $V_{Ox}$, $V_{Oy}$, the relative coordinate x(t) and y(t) can be generated as

$$x(t) = x_T(t) - x_O(t), \tag{14}$$

$$y(t) = y_T(t) - y_O(t). \tag{15}$$

Define state variables in mixed coordinates (i.e., a combination of polar and rectangular) as

$$x_1(t) = \beta(t) \ , \tag{16}$$

$$x_2(t) = r(t), \tag{17}$$

$$x_3(t) = V_{Tx}(t) - V_{Ox}(t) = V_x(t), \tag{18}$$

$$x_4(t) = V_{Ty}(t) - V_{Oy}(t) = V_y(t), \tag{19}$$

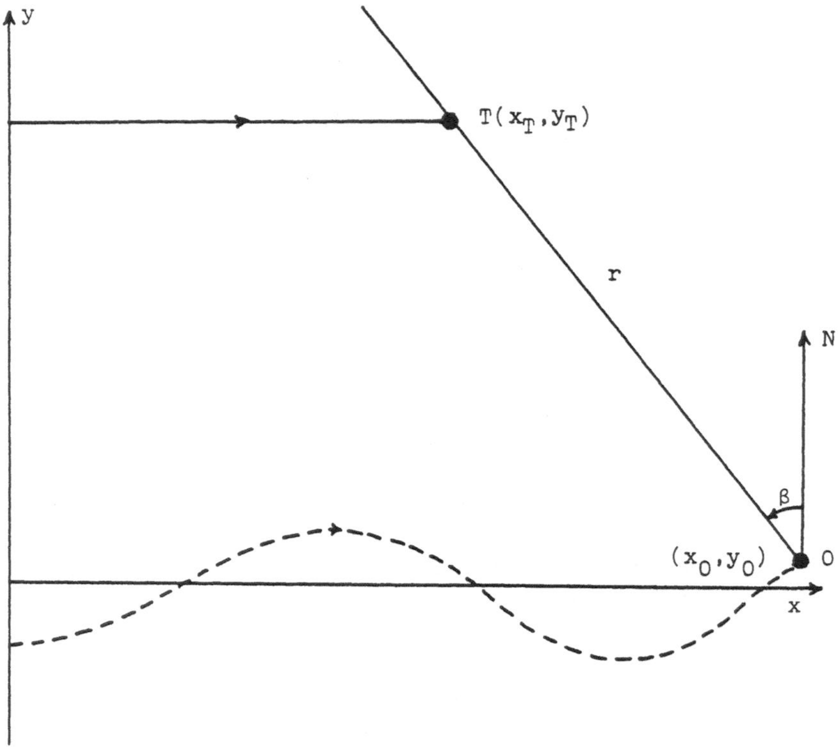


Figure 1. BOT configuration

where $\beta(t)$ is bearing of T from O and r is range. Then, the state equation in this specific coordinate system becomes

$$\dot{x}(t) = \begin{bmatrix} \dfrac{x_3 \cos x_1 - x_4 \sin x_1}{x_2} \\ x_3 \sin x_1 + x_4 \cos x_1 \\ a_x \\ a_y \end{bmatrix}, \tag{20}$$

where $a_x(t)$, $a_y(t)$ are acceleration in each direction. Measurement of $\beta(t)$ leads to

$$y(t) = [\,1 \quad 0 \quad 0 \quad 0\,]\; x(t). \tag{21}$$

Observability is checked next for the two cases where maneuvering exists, i.e., $a_x(t) \neq 0$ and/or $a_y(t) \neq 0$, and nonmaneuvering, i.e., both are zero. From (20) and (21) with $a_x(t) = 0$ and $a_y(t) = a(t) \neq 0$, (i.e., maneuvering exists only in one direction), and by successive replacement

$$y = x_1, \tag{22}$$

$$y' = \frac{x_3 \cos y - x_4 \sin y}{x_2}, \tag{23}$$

$$y'' = \frac{-(a \sin y + 2y'x_4 \cos y + 2y'x_3 \sin y)}{x_2}, \tag{24}$$

$$y''' = \frac{3ay'\cos y + x_3(3y''\sin y + 2(y')^2\cos y) + x_4(3y''\cos y - 2(y')^2 \sin y) + a'\sin y}{x_2}. \tag{25}$$

Then, from (22) - (25),

$$x_1 = y, \tag{26}$$

$$x_2 = \frac{-2y'x_4 - a \cos y \cdot \sin y}{y'' \cos y + 2(y')^2 \sin y}, \tag{27}$$

$$x_3 = \frac{(y'' \sin y - 2(y')^2 \cos y)\, x_4 - y'a \sin y}{y'' \cos y + 2(y')^2 \sin y}, \tag{28}$$

$$x_4 = \frac{a[4(y')^3\cos y \sin y + 6y'y''\cos^2 y - 3y'y'' - y'''\cos y \sin y] + a' \sin y\, [y'' \cos y + 2(y')^2 \sin y]}{2y'y''' - 3(y'')^2 + 4(y')^4}, \tag{29}$$

From (29) it is clear that it $a(t)$ and/or $a'(t) \neq 0$, (i.e., maneuvering exists), $x_4$ is connected to the measurement vector Y, it is unique, and thus it is observable. This implies from (27) and (28) that $x_2$ and $x_3$ are also uniquely connected. So, the system satisfies the connected condition in this case. But when $a(t) = 0$ and $a'(t) = 0$, i.e. nonmaneuvering, (29) suggests that $x_4$ is not connected to Y and is unobservable. This causes, again from (27) and (28), that $x_2$ and $x_3$ are disconnected from Y and thus unobservable also. Only $x_1$ is observable which is itself a measurement variable. After lengthy computation, the determinant of the Jacobian becomes

$$\det J = \frac{-2y'a'\sin y+3a[2(y')^2\cos y+y''\sin y]-[12y'y''\sin y(1+\cos^2 y)}{x_2^4}$$

$$\frac{+\,8\cos^3 y(y')^3]\,x_3+4y'\cos y\,\sin y[2(y')^2\cos y+3y''\sin y]x_4}{} \; . \qquad (30)$$

From (30), this system is unobservable with det J = 0 for the following cases (among others):

1) Infinite range, $x_2 = \infty$;
2) Zero heading rate and acceleration, $\beta' = \beta'' = 0$;
3) $X_3 = X_4 = 0$, i.e., $\dot{x}_2 = 0$, with $a(t) = a'(t) = 0$ (parallel stationary movement, including tail chase);
4) Constant range with special heading such that

$$\tan\beta = \frac{fa(\beta')^2}{2\,a'\beta' - 3a\beta''} \; . \qquad (31)$$

As well as certain others, the system is unobservable due to the lack of rank and thus, lack of information of some states in those cases. Consequently, the known result, i.e., bearing-only target tracking system, is observable when relative maneuvering exists with several exceptional conditions such as described in 1) through 4) above.

Example 3, [22]

Another important application area of the system observability in the ocean environment is the SONAR tracking where a number of sensors deployed and measurement policy are changed. Here, one is interested in the determination of the number of sensors which give enough information to estimate all of the states with required accuracy and determination of the measurement scheme which yields the maximum information if the number of sensors available are limited.

Figure 2 shows sensor and target configuration for up to three sensors in a vertically deployed line array.

In a two-dimensional space, any point target is described by the four states, position, and velocity in each direction. Since sound speed varies with depth, salinity, and temperature, it is also included in the state space, i.e., define the state variables as follows:

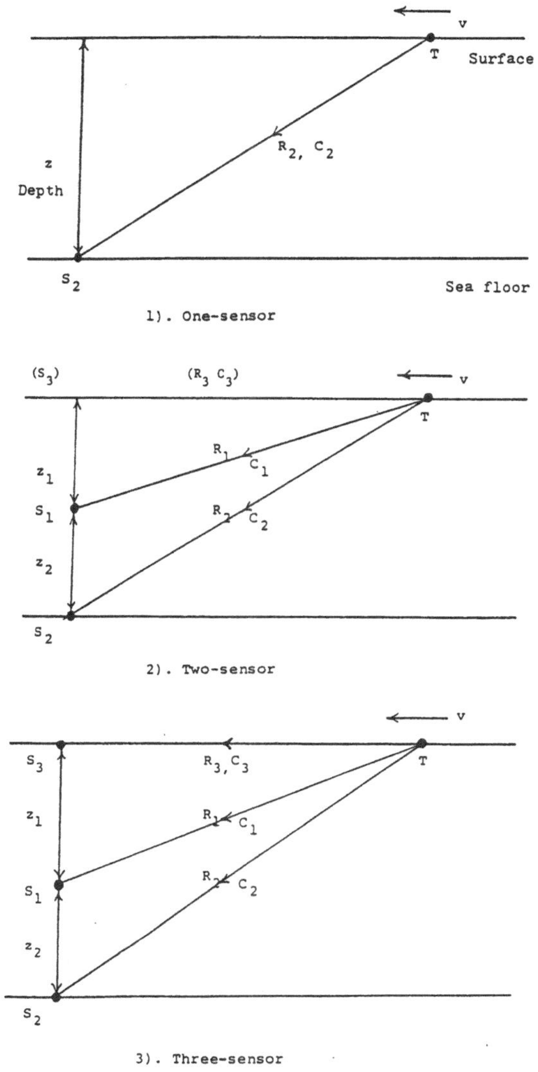


Figure 2. Sensor configuration.

$x_1(t)$ is target position in x-direction;

$x_2(t)$ is target velocity in x-direction;

$x_3(t)$ is target position in Y-direction;

$x_4(t)$ is target velocity in Y-direction;

$x_5(t)$ is $C_1$ (Acoustic wave speed in $R_1$);

$x_6(t)$ is $C_2$ (Acoustive wave speed in $R_2$).

Then the system equation can be written simply as

$$\dot{x}_1 = x_2 ,$$
$$\dot{x}_3 = x_4 , \tag{32}$$
$$\dot{x}_2 = \dot{x}_4 = \dot{x}_5 = \dot{x}_6 = 0 .$$

Measurement quantities may be absolute time delay or Doppler shift between sensor and target (synchronized or active SONAR case), or relative delay or Doppler shift between two sensors (active or passive).

To observe deterministic observability for this system, categorize the measurement scheme into three groups for convenience as:

a) an absolute delay; 1S1D, (i.e., 1 sensor, 1 delay)

b) pure relative delay; 2S1D, 3S2D, 3S3D, (2 sensors, 1 delay, and so forth)

c) relative Doppler; 2S1P, 2S1D1P, 3S2D1P, (2 sensors, 1 Doppler, and so forth).

The first case when an absolute time propagation delay of the acoustic wave with one-sensor deployment at $S_2$ gives the observation equation

$$y(t) = \frac{R_2(t)}{C_2(t)} . \tag{33}$$

Considering system equation (32) and the relation with omission of t

$$R_2 = (x_1{}^2 + x_3{}^2)^{1/2} ,$$

$$\dot{R}_2 = \frac{x_1x_2 + x_3x_4}{R_2} . \quad (34)$$

Then, by algebraic manipulation, it is clear even without solving for x that $x_5$ does not appear in any equation explicitly. So $x_5$ is not connected to the measurement vector Y, with

$$Y = \{ y, y', \ldots, y^{(5)} \} .$$

It is unobservable and makes the system unobservable deterministically. Actual solution of those equations shows that the remaining variables have multiple solutions, i.e., they are observable only in a wide sense. Here one can consider due to physical argument, the possibility of dropping certain multiple solutions which may appear in the global inverse domain. This is similar to adding a restriction on the inverse (solution) domain, i.e., making a regionally inverse instead of a global one. So, deleting any physically impossible solutions means shrinkage of the observability concept to a regional sense from a global one.

In the second case when pure relative delay is measured, for example as in 2S1D, then

$$\begin{aligned} y &= \tau_{12} , \\ &= \frac{R_2}{C_2} - \frac{R_1}{C_1} , \\ &= \frac{(x_1{}^2 + x_3{}^2)^{1/2}}{x_6} - \frac{(x_1{}^2 + (x_3 - z_2)^2)^{1/2}}{x_5} , \end{aligned} \quad (35)$$

$$y' = \frac{x_1x_2 + x_3x_4}{x_6R_2} - \frac{x_1x_2 + (x_3 - z_2)\, x_4}{x_5R_1} . \quad (36)$$

Continuation up to (n-1)th order shows that the results are almost identical to the first case except $x_5$ appears in the expressions. It implies immediately that the process, as well as individual states, is observable at least in a wide sense. When adding more sensors like 3S2D or 3S3D, the system becomes more observable due to the increasing possibility of uniqueness of the solution in terms of state x.

In the last case, when the measurement equation includes Doppler shift as in 2S1P, 2S1D1P, or 3S2D1P, an interesting case results. For example, when one Doppler shift in a two-sensor deployment (2S1P) is observed,

$$y = f_{12} ,$$

$$= f_c\left(\frac{\dot{R}_2}{c_2} - \frac{\dot{R}_1}{c_1}\right) ,$$

$$= f_c \left(\frac{x_1x_2 + x_3x_4}{x_6R_2} - \frac{x_1x_2 + (x_3 - Z_2)\, x_4}{x_5R_1}\right) ,$$

$$= f_c\, y_D' , \qquad (37)$$

where $y_D'$ is relative delay measurement differentiation (36) and $f_c$ is carrier frequency of the system. Continuation yields

$$y' = f_c\, y''_D ,$$

$$y'' = f_c\, y'''_D ,$$

$$\vdots$$

$$y^{(5)} = f_c\, y_D^{(6)} . \qquad (38)$$

Thus, Doppler measurement is similar to scaling up of delay difference with scaling factor $f_c$. But, as discussed before, the 2S1D system itself is already observable (at least in a wide sense). So, this system is also observable in the same context. The same argument can be applied for the 2S1D1P or 3S2D1P measurement cases. Thus, the Doppler measurement system is observable deterministically as far as the delay measurement system is observable. Of course, the scaling factor influences the magnitude of the information obtained from the measurement. The effect of this will be seen in the following stochastic observability analysis.

3. STOCHASTIC "OBSERVABILITY" USING INFORMATION THEORY

Contrary to the deterministic system, indeed, the meaning of the term "observability" is not so clearcut here. Obviously, an unobservable BOT or array SONAR system discussed earlier may be observable in some sense if this system if driven by a random process (at least theoretically). Of course, this noise may otherwise have a reverse effect on any state tracking.

Here, a relative sense of stochastic system observability is developed and used as a criterion based on well-established information theory, i.e., the amount of information about one random function contained in another random function, so called, mutual information between state $x_t$ and observation $y_t$, is used as a criterion to determine degree of observability. Consider a general form of stochastic nonlinear process of Ito type,

$$dx_t = f(x_t)\,dt + g(x_t)\,dw_t,\quad x_{to} = x_o, \tag{39}$$

$$dy_t = h(x_t)\,dt + dv_t, \tag{40}$$

where $x_t \epsilon R^n$, $y_t \epsilon R^m$, $f(\cdot)$, $g(\cdot)$ and $h(\cdot)$ are appropriate functions of their arguments. Noise processes $\{w_t\}$ and $\{v_t\}$ are assumed to be Wiener processes independent of each other and having unit variance. Also, $x_o$ is assumed independent of those noises.

For the conditional density $P_t(a)$ where $P_t(a) = dF(x_t < a|Y_s)/da$, $Y_s$, a sub-$\sigma$-algebra generated by $\{y_s,\ s<t\}$, the Fisher quantity of information $J(x_t, y_t)$ is

$$J(x_t, y_t) = -E\left[\frac{\partial^2}{\partial a \partial a^T} \ln P_t(a)\right], \tag{41}$$

and the corresponding shannon quantity $I(x_t, y_t)$ is

$$I(x_t, y_t) = E\left[\ln \frac{P_t(a)}{\rho_t(a)}\right], \tag{42}$$

where $\rho_t(a) = dF(x_t < a)/da$ is an unconditional density for the process $\{x_t\}$ from (39). Of course, $P_t(a)$ and $\rho_t(a)$ satisfy Kolmogorov's forward equation and Kushner's equation, respectively. Due to the functional dependence of $f(\cdot)$ and $g(\cdot)$, a state $x_t$, derivation of any general relationship between the two information concepts in (41), (42) are generally very difficult. If, at least, the system dynamic equation (39) is linear in state $x_t$ with $g(t)$ only a function of t, then the Fisher and Shannon information can be related as given in [15].

Comparison of the definitions show that Fisher information traditionally represents the information which is contained in a single density (i.e., either a priori or a posteriori density), but Shannon's mutual information compares the system information difference with and without observation.

Traditionally, entropy of the process $x_t$ is defined as

$$H(x_t) = -E[\ln \rho_t(x_t)], \tag{43}$$

where $\rho_t(x_t)$ as in (42).

Similarly, conditional entropy is defined as

$$H(x_t|y_t) = -E[\ln P_t(x_t)]. \tag{44}$$

Note that $P_t(x_t)$ is a conditional density. Then, from the definition (42)

$$I(x_t, y_t) = H(x_t) - H(x_t|y_t). \tag{45}$$

So, to compute (45) marginal entropy $H(x_t)$ and conditional entropy $H(x_t|y_t)$ may be computed. For this purpose, the usual entropy-variance relation

$$H(x_t) = \frac{n}{2}\ln A + \frac{1}{2}\ln(\det \Gamma_{Tt}), \tag{46}$$

$$H(x_t|y_t) = \frac{n}{2}\ln A + \frac{1}{2}\ln(\det P_{Tt}), \tag{47}$$

can be used. (46), (47) hold for a large class of densities $\rho_t(x_t)$ and $P_t(x_t)$ if their means and covariances exist. Here A is a constant determined by the densities, and $\Gamma_{Tt}$ is the covariance of the density $\rho_t(x_t)$. $P_{Tt}$ is the covariance of the density $P_t(x_t)$.

Then (47) becomes

$$I(x_t, y_t) = H(x_t) - H(x_t|y_t) = \frac{1}{2}\ln\left(\frac{\det \Gamma_{Tt}}{\det P_{Tt}}\right), \quad \Gamma_{T0} = P_{T0}. \tag{48}$$

It should be noted that this relationship holds for not only Gaussian processes but most others of interest. By setting $\Gamma_{T0} = P_{T0}$, the same initial information at $t_0$, initial observability always starts from zero. This gives an effect of normalization. Equation (48) is used as a measure of state observability. For the individual i-th state,

$$I_i(x_i, y_t) = \frac{1}{2} \ln \left(\frac{\Gamma_{Tii}}{P_{Tii}}\right), \qquad i = 1, 2, \ldots, n . \tag{49}$$

With some state estimation approximation, the resultant mean $\hat{x}_t = E[x_t | Y_s, 0 < s < t]$ and covariances $P_t$, $\Gamma_t$, (49) becomes

$$I(\hat{x}_t, y_t) = H(\hat{x}_t) - H(\hat{x}_t | y_t) = \frac{1}{2} \ln\left(\frac{\det \Gamma_t}{\det P_t}\right), \qquad \Gamma_0 = P_0 . \tag{50}$$

So, in this case the degree of observability at any time t may be estimated by the quantities $P_t$ and $\Gamma_t$.

An approximation method of computing $P_t$ and $\Gamma_t$ is

$$\dot{\Gamma}_t = f_x(\bar{x}_t)\, \Gamma_t + \Gamma_t\, f_x^T(\bar{x}_t) + g(t)\, Q_t g^T(t), \qquad \Gamma_0 = P_0 . \tag{51}$$

Here

$$f_x(\cdot) = \frac{\partial f(\cdot)}{\partial x_t}\Big|_{x_t = \bar{x}_t} ,$$

$\bar{x}_t = E[x_t | X_t]$, $X_t$ is the sub-$\sigma$-algebra generated by $\{x_s\}$, $0 < s < t$,

and

$$\dot{P}_t = f_x P_t + P_t f_x^T + g(t)\, Q_t g^T(t) - P_t h_x^T R_t^T h_x P_t, \quad P_{to} = P_o, \tag{52}$$

where

$$h_x(\cdot) = \frac{\partial h(\cdot)}{\partial x_t}\Big|_{x_t = \hat{x}_t} .$$

Both in (51) and (52), it is assumed that $g(\cdot)$ is only a function of time t.

## 4. INFORMATION STRUCTURE ANALYSIS OF BOT AND LINEAR ARRAY SONAR

### 4.1 BOT System

Observability is an important consideration in underwater SONAR signal processing which was discussed above. Here the same problem is discussed in a stochastic sense. The effect of deterministic observability on stochastic "observability" is studied with simulation using the known fact that the BOT system is observable when proper relative maneuvering exists. The three different coordinate systems (rectangular, modified polar, and mixed) are adopted as in Table 1. According

Table I. System description of different coordinates.

| | Rectangular | Modified polar | Mixed |
|---|---|---|---|
| State variable | $x(t)=\begin{bmatrix}x_1\\x_2\\x_3\\x_4\end{bmatrix}=\begin{bmatrix}r_x\\v_x\\r_y\\v_y\end{bmatrix}$ | $x(t)=\begin{bmatrix}x_1\\x_2\\x_3\\x_4\end{bmatrix}=\begin{bmatrix}\dot{B}\\\dot{r}/r\\B\\1/r\end{bmatrix}$ | $x(t)=\begin{bmatrix}x_1\\x_2\\x_3\\x_4\end{bmatrix}=\begin{bmatrix}B\\r\\v_x\\v_y\end{bmatrix}$ |
| State eqs. | $\dot{x}(t)=\begin{bmatrix}0&1&0&0\\0&0&0&0\\0&0&0&1\\0&0&0&0\end{bmatrix}x(t)+ a_y$ $+w(t)$ | $\dot{x}(t)=\begin{bmatrix}-x_1x_2-a_yx_4\sin(x_3)\\x_1^2-x_2^2+a_yx_4\cos(x_3)\\x_1\\-x_2x_4\end{bmatrix}$ $+\ w(t)$ | $\dot{x}(t)=\begin{bmatrix}\frac{x_3\cos(x_1)-x_4\sin(x_1)}{x_2}\\x_3\sin(x_1)+x_4\cos(x_1)\\0\\a_y\end{bmatrix}$ $+\ w(t)$ |
| Meas. eqs. | $y_k=h(x_k,k)+v_k$ $=\tan^{-1}(\frac{x_1}{x_3})+v_k$ | $y_k=H\ x_k+v_k$ $=[\ 0\ 0\ 1\ 0\ ]\ x_k+v_k$ | $y_k=H\ x_k+v_k$ $=[1\ 0\ 0\ 0\ ]\ x_k+v_k$ |

to the physical process, a continuous system, discrete-measurement algorithm for a truncated, second-order filter is implemented in two stages. At the first measurement update, the observed data is processed through the discrete filter. At the second, time-propagation stage, time integration of the first and second moments is made according to the continuous case.

With the configuration shown in Figure 1, maneuvering is given as

$$a_x(t) = 0,$$

$$a_y(t) = -0.025\cos(0.005\ t), \qquad m/s^2 .$$

Initial states are assumed Gaussian with appropriate mean. Other parameters used are

$T$ = 10 sec (sampling interval; can be changed),

$\Delta t$ = 1 sec (time update interval),

$r(0)$ = 8000 m (initial range),

$V_{Tx}$ = 10 m/s ≈ 20 KTs (target velocity in x-direction),

$V_{Ty}$ = 0,

$V_{0x}$ = 15 m/s ≈ 30 KTs (observer velocity in x-direction),

$V_{0y}$ = 5 sin (0.005t) m/s.

The measurement noise sequence and system noise are also assumed to be Gaussian with covariance $R_k$ and $Q_t$, respectively. With the algorithm and parameters as above, a simulation is made for the three coordinates and changing certain important variables.

As above, the BOT is deterministically unobservable when no relative maneuvering exists, and observable when nontrivial maneuvering exists.

Figures 3, 4, and 5 show the degree of observability in range error which is affected by changing maneuvering $a_y$ and system noise $Q_t$. For all three coordinates, (mixed, rectangular, and modified polar), range variable is strongly "observable" when $a_y \neq 0$ and $Q_t = 0$. But it becomes very weakly "observable" when $a_y = 0$. System noise impairs both range and total "observability" strongly. Range error does not decrease in any sense when there is no relative maneuvering. Typical information structure is shown in Table II (mixed case). Most strong "observability" is presented by the observed variable $\beta(t)$ in this case.

The estimated information structure, "observability", is strongly affected by the measurement-noise covariance $R_K$. As $R_K$ increases, for example, from $(0.2°)^2$ to $(6°)^2$, observability becomes poor. Mixed coordinates show the least effect; thus the least error for increased noise level. The modified-polar case showed instability in the range variable due to the poor information in the early stage.

The sampling interval is another parameter which affects the information structure. More frequent sampling, of course, gives stronger observability.

## 4.2 Linear Array SONAR

For this system, seven possible measurement policies are chosen with the linear deployment of up to three sensors. Here, the popular

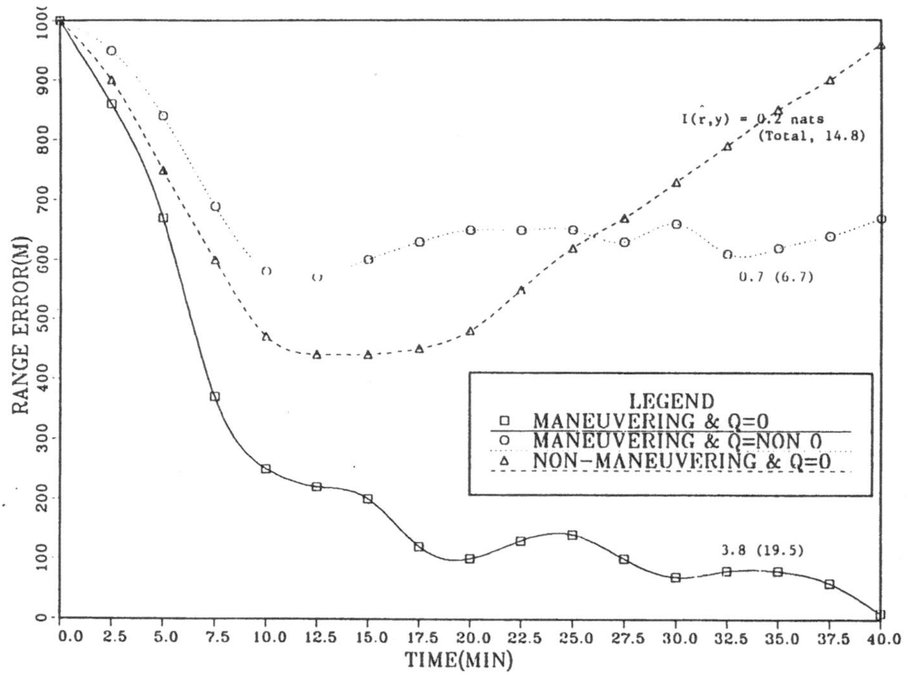


Figure 3. OBS. & Range Error (Mixed).

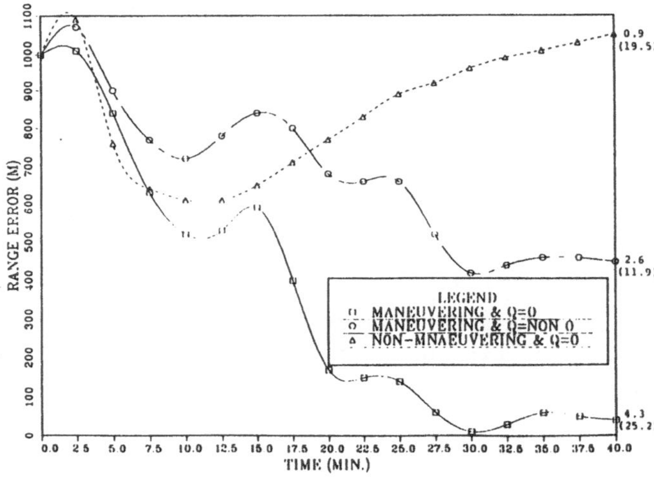


Figure 4. OBS. & Range Error (Rec.).

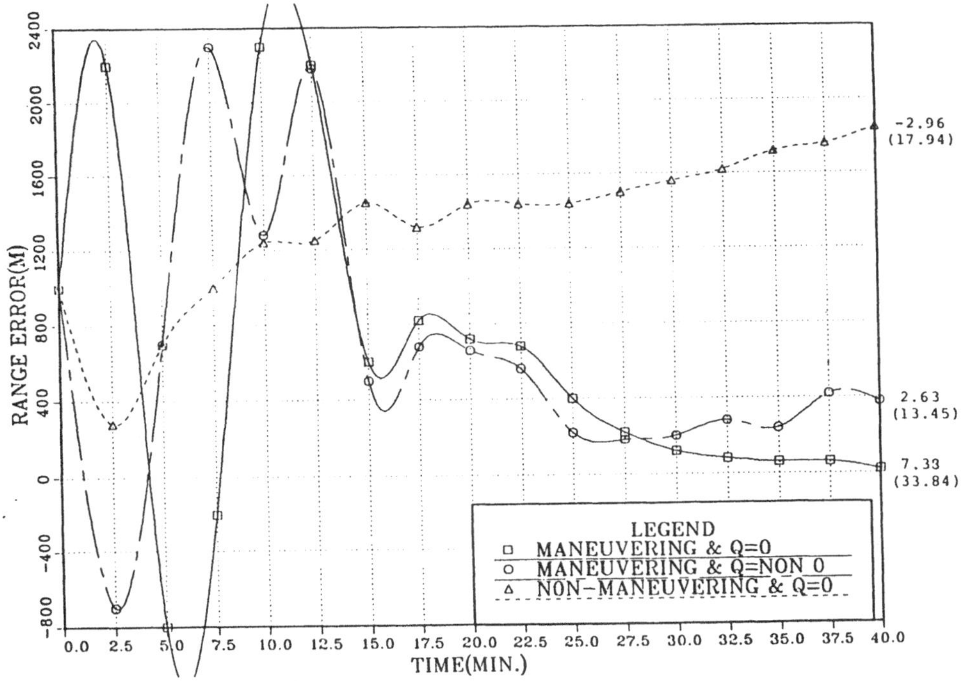


Figure 5. OBS. & Range Error (MP).

Table II. Mix-Coordinate Information Estimate, Nats

| | Q = 0 | | | | | | | | | | Q ≠ 0 | | | | |
|---|---|---|---|---|---|---|---|---|---|---|---|---|---|---|---|
| | $A_y \neq 0$ | | | | | $A_y = 0$ | | | | | $A_y \neq 0$ | | | | |
| t(min) | Tot. Obs. | β | r | $V_x$ | $V_y$ | Tot | β | r | $V_x$ | $V_y$ | Tot | β | r | $V_x$ | $V_y$ |
| 0 | 0 | 0 | 0 | 0 | 0 | 0 | 0 | 0 | 0 | 0 | 0 | 0 | 0 | 0 | 0 |
| 2.5 | 4.5 | 2.8 | 0.8 | 0.3 | 0 | 4.4 | 2.7 | 0.8 | 0.3 | 0 | 2.8 | 2.5 | 0.5 | 0.2 | 0 |
| 5 | 7.3 | 4.6 | 1.3 | 0.3 | 0 | 7.0 | 4.5 | 1.1 | 0.4 | 0 | 4.1 | 3.8 | 0.8 | 0.2 | 0 |
| 10 | 10.4 | 6.8 | 1.9 | 0.7 | 0.01 | 9.6 | 6.3 | 1.0 | 0.4 | 0.01 | 5.8 | 5.3 | 1.1 | 0.2 | 0.01 |
| 15 | 12.8 | 6.9 | 2.0 | 1.1 | 0.1 | 11.2 | 7.3 | 0.9 | 0.4 | 0.04 | 5.8 | 5.5 | 1.0 | 0.4 | 0.03 |
| 20 | 15.2 | 7.3 | 3.1 | 2.7 | 0.1 | 12.3 | 7.9 | 0.7 | 0.5 | 0.1 | 6.2 | 5.5 | 0.8 | 0.5 | 0.04 |
| 25 | 16.3 | 8.1 | 3.0 | 3.0 | 0.2 | 13.1 | 8.2 | 0.5 | 0.5 | 0.3 | 6.4 | 5.7 | 0.7 | 0.5 | 0.1 |
| 30 | 17.8 | 7.6 | 3.4 | 3.5 | 1.0 | 13.8 | 8.3 | 0.3 | 0.5 | 0.6 | 6.8 | 5.9 | 0.7 | 0.6 | 0.1 |
| 35 | 18.8 | 8.6 | 3.7 | 3.9 | 1.4 | 14.4 | 8.4 | 0.2 | 0.5 | 1.0 | 6.7 | 5.8 | 0.6 | 0.6 | 0.2 |
| 40 | 19.5 | 8.2 | 3.8 | 4.0 | 1.7 | 14.8 | 8.4 | 0.2 | 0.5 | 1.2 | 6.7 | 5.7 | 0.7 | 0.7 | 0.2 |

$R = (1°)^2$, $T = 10$ s.

extended Kalman filter of the discrete type is used. Other parameters are as follows:

Measurement sampling interval T = 15 sec, and initial condition of the estimate (when no initial error is assumed) $\hat{x}(0)$ with components

$\hat{x}_1(0) = 10{,}000$ m ,

$\hat{x}_2(0) = -15.433$ m/s ($\approx$ 30 knots, approaching the sensor ,

$\hat{x}_3(0) = 4000$ m ,

$\hat{x}_4(0) = 0$ m/s ,

$\hat{x}_5(0) = 1500$ m/s ,

$\hat{x}_6(0) = 1500$ m/s .

Gaussian state noise variances are

$\sigma_1 = \sigma_x = 100$ m,

$\sigma_2 = \sigma_{Vx} = 0.15$ m/s,

$\sigma_3 = \sigma_y = 40$ m,

$\sigma_4 = \sigma_{Vy} = 0.1$ m/s,

$\sigma_5 = \sigma_{c_1} = 5$ m/s,

$\sigma_6 = \sigma_{c_2} = 5$ m/s,

and means are zero.

The measurement noises are zero mean white Gaussian sequence with covariance

$\sigma_{\tau_{12}} = 0.019$ sec,

$\sigma_{\tau_{23}} = 0.026$ sec,

$\sigma_{\tau_{13}} = 0.016$ sec,

$\sigma_{abs.D} = 0.359$ sec,

$\sigma_{f_{12}} = 0.1875$ Hz .

$$P(0) = \begin{bmatrix} \sigma_x^2 \times 10^4 & & & & & \\ & \sigma_{V_x}^2 \times 5 \times 10^2 & & & & \\ & & \sigma_y^2 \times 10^4 & & & \\ & & & \sigma_{Vy}^2 \times 5 \times 10^2 & & \\ & & & & \sigma_{c_1} \times 10^2 & \\ & & & & & \sigma_{c_2} \times 10^2 \end{bmatrix}$$

$= \Gamma(0)$,

and

$f_c = 3500$ Hz (carrier frequency of modulation),

$z_1 = 2000$ m (intersensor distance of $s_1$ and $s_3$),

$z_2 = 2000$ m (intersensor distance of $s_1$ and $s_2$).

With the above parameters, 20 runs were averaged.

Table III shows the estimated total mutual information content for the various measurement schemes. Deployment of an increased

**Table III. Total Observability, Nats**

| # of sensors & Meas. / Time (min.) | 1S (Abs.D) | 2S (1D) | 2S (1P) | 2S (1D1P) | 3S (2D) | 3S (3D) | 3S (2D1P) |
|---|---|---|---|---|---|---|---|
| 0 | 0.0 | 0.0 | 0.0 | 0.0 | 0.0 | 0.0 | 0.0 |
| 0.5 | 5.0 | 6.1 | 7.2 | 13.5 | 13.4 | 14.0 | 21.7 |
| 1.0 | 5.7 | 7.6 | 8.6 | 15.5 | 16.3 | 17.9 | 24.7 |
| 1.5 | 6.1 | 8.2 | 9.5 | 16.9 | 17.3 | 18.8 | 26.1 |
| 2.0 | 7.0 | 8.6 | 10.1 | 17.8 | 17.9 | 19.4 | 27.1 |
| 2.5 | 8.1 | 8.9 | 10.6 | 18.4 | 18.2 | 19.7 | 27.8 |
| 3.0 | 8.6 | 9.2 | 11.3 | 18.9 | 18.4 | 20.0 | 28.5 |
| 3.5 | 8.8 | 9.7 | 11.7 | 19.4 | 18.8 | 20.3 | 29.1 |
| 4.0 | 8.9 | 9.9 | 12.4 | 19.8 | 19.1 | 20.6 | 29.7 |
| 4.5 | 9.1 | 10.2 | 12.9 | 20.2 | 19.4 | 20.9 | 30.2 |
| 5.0 | 9.2 | 10.5 | 13.4 | 20.6 | 19.7 | 21.2 | 30.8 |

number of sensors seems to yield stronger observability. In spite of large measurement magnitude, 1S1 abs.D system shows the weakest observability due to the unobservable state variable $x_5 (= C_1)$ which also is indicated earlier in the deterministic observability analysis. With only one delay or one Doppler with two-sensor deployment (2S1D or 2S1P), the system remains weakly "observable", even though both cases are deterministically observable. Information estimate is improved when both delay and Doppler are measured (2S1D1P) or when one more sensor is added with the measurement of only one quantity, delay here. Estimated information does not increase appreciably by adding the same kind of measurement quantity as can be seen from 3S2D and 3S3D cases. This may be caused by the fact that the third delay measurement depends on the first two delays. Only two delays are independent in the three-sensor delay measurement.

Strongest information increase is achieved when one observes both delay and Doppler with three sensors (3S2D1P; 30.8). It is also of interest that most of the information is obtained during the very early stage, i.e., when the first few measurement data are processed.

Information analysis of Table IV show that "observability" of $v_x$, $v_y$, and $C_1$, $C_2$ are relatively poor compared to the range variables $r_x$, $r_y$. The effect of filtering error due to different degrees of information content is estimated by Figure 6 for range $v_x$, as an example.

Table IV. Estimated Information Contents, Nats (at final time t = 5 min)

| Meas. / State | 1S1D | 2S1D | 2S1P | 2S1D1P | 3S2D | 3S3D | 3S2D1P |
|---|---|---|---|---|---|---|---|
| $r_x$ | 4.0 | 2.2 | 3.5 | 3.5 | 2.8 | 3.1 | 4.6 |
| $V_x$ | 0.2 | 0.2 | 0.8 | 0.9 | 0.2 | 0.3 | 1.1 |
| $r_y$ | 2.3 | 1.5 | 1.8 | 2.4 | 3.2 | 3.4 | 4.2 |
| $V_y$ | 0.1 | 0.3 | 0.2 | 1.1 | 1.4 | 1.5 | 2.2 |
| $C_1$ | 0.0 | 0.5 | 0.6 | 0.7 | 2.0 | 2.1 | 2.4 |
| $C_2$ | 0.05 | 0.6 | 0.5 | 0.7 | 1.8 | 1.9 | 2.2 |
| Total | 9.2 | 10.5 | 13.4 | 20.6 | 19.7 | 21.2 | 30.8 |

Increased measured quantities with more sensors yield less filtering error due to improved "observability". This trend seems quite natural intuitively. With initially given 1000 m range error, combinational measurement of Doppler and delay yields significantly small errors. The 3S2D1P case, especially, shows very desirable characteristics (see Figure 6). It is important to note here that very "undesirable" (refers to large errors or oscillation of range error) result is shown when only time delay is measured. Figure 6 shows large (more than initial error) errors in case of 2S1D, 3S3D and some overshoot appears for 1S1D even with reasonably good range information. So, it seems that Doppler measurement is crucial for good SONAR range estimation when it is combined with delay measurements in such linear configurations.

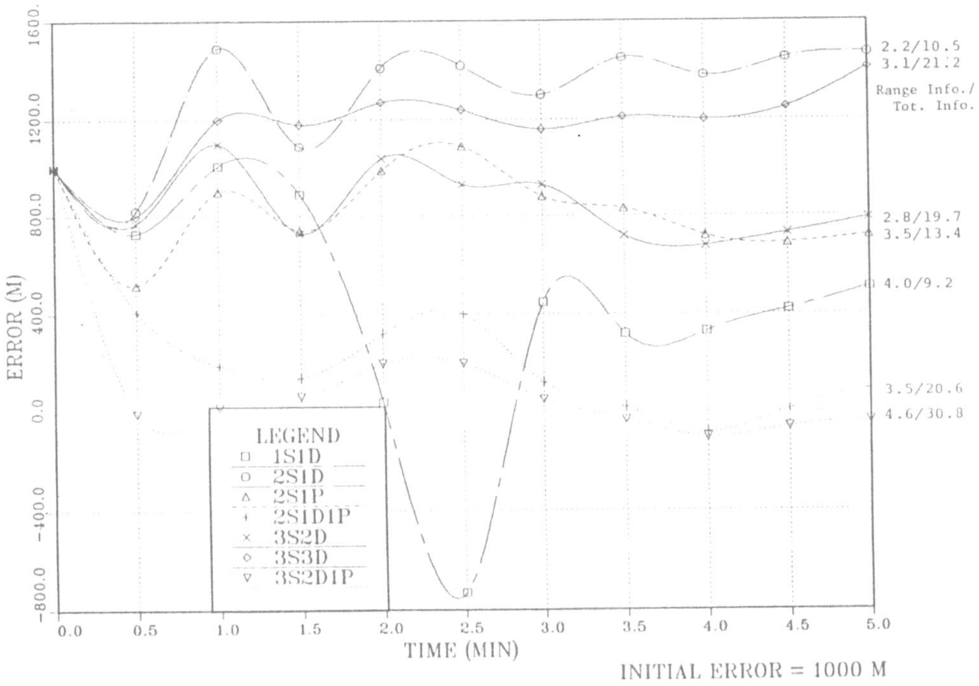


Figure 6. Range Error.

In general, simulation indicates that the 1S1D, 2S1D, and 2S1P group shows the poorest performance; the 2S1D1P, 3S2D, and 3S3D group shows medium range error, and 3S2D1P shows superior performance. The 1S1D measurement shows, also, a couple of overshoots with the poorest information.

For comparison, the discrete version of Fisher information matrix J(k,1) is computed by

$$J(k,1) \triangleq \sum_{i=1}^{k} \phi^{-T}(k,i)\; H^T(\hat{x}(i))\; R(i)^{-1} H(\hat{x}(i))\; \phi^{-1}(k,i)\;, \quad (53)$$

or, its iterative modification,

$$J(k+1,1) \triangleq \phi^{-T}(k+1,k)\; J(k,1)\; \phi^{-1}(k+1,k)$$

$$+ H^T(\hat{x}(k+1))\; R^{-1}(k+1)\; H(\hat{x}(k+1)). \quad (54)$$

The selected five observation policies are computed and shown in table V. Matrix J(k,1) remains singular for the entire period of observation for the 1S1D measurement case and nonsingular with increasing determinants with increasing number of sensors. Comparison of Table V with the total information content (Table III) reveals that the two approaches exactly coincide with the same order of measurement policy observability.

Table V. Fisher Information for Linear Array.

| Meas. Time | 1s(1D) | 2s(1D1P) | 3s(2D) | 3s(3D) | 3s(2D1p) |
|---|---|---|---|---|---|
| 0 min. | | | | | $5.33 \times 10^{-17}$ |
| 0.5 | | $4.88 \times 10^{-21}$ | $8.13 \times 10^{-22}$ | $3.8 \times 10^{-21}$ | |
| 2.5 | Unobservable | $3.98 \times 10^{-15}$ | $2.56 \times 10^{-11}$ | $2.0 \times 10^{-10}$ | $1.08 \times 10^{-9}$ |
| 5.0 | | $3.73 \times 10^{-9}$ | $7.63 \times 10^{-8}$ | $6.3 \times 10^{-7}$ | $1.24 \times 10^{-5}$ |

Superiority of the measurement 3S2D1P system is also indicated for both methods.

## 5. CONCLUSION

Data (state) observability for both deterministic and stochastic cases are studied here.

Since nonlinear observability for deterministic systems is a geometric nonlinear functional structure property, the inverse function theorem is useful. The nonzero Jacobian condition, which can be related to n-1 dimensions for the special case, provides the connectedness condition for every state to be connected to the measurements. However, a finite-covering condition must be tightened to a one-covering condition, then by which univalence of the connection can be guaranteed.

Depending on the satisfaction of these two conditions, observability in the strict sense, observability in the wide sense, and unobservable states are determined. Application of this method is demonstrated by the two examples, BOT and linear-array SONAR. A well-known fact that the BOT system is observable when relative maneuvering exists and unobservable when nonmaneuvering is proven. At least two sensors are necessary for the linear, array-SONAR observability. Doppler measurement scales up the delay measurement by the factor of carrier frequency.

Instead of using the Fisher information matrix, mutual information (Shannon sense) is utilized to determine the degree of "observability" of the stochastic process. In the latter approach, the common information between the state and observed data is computed, i.e., the information content of the state $x_t$ which is contained in the measurement data is estimated by proper filter algorithm.

Obvious advantages of this approach over the Fisher-information approach in connection with linearized approximations, such as through the extended Kalman filter, are as follows:

1) Total and individual observable state information can be computed even in the case that some states are unobservable. This is not possible in the Fisher information approach due to the singularity of the observability matrix.
2) Identification of unobservable states in the current approach is also immediate by just inspection of the state where information is not growing. But this is very difficult in the Fisher information approach where they can only be identified by empirical guessing or trial and error [23].
3) Fisher information, while still valid basically for the nonlinear system, is only useful for first-order linearization if computed by the normal, convenient equation (53). But Shannon information is estimated with any higher order approximation.
4) Fisher information, in the conventional sense, accommodates only measurement noise. But Shannon information readily considers both system and measurement noise effects.

The information analysis for BOT system shows that both speed and range are weakly observable when the observer does not maneuver relative to the target for all three considered coordinates. Corresponding errors are, naturally, big and do not decrease in any sense. But when there is relative maneuver, information grows continuously for all states, thus making the process strongly observable. Error decreases very fast accordingly.

For the linear array SONAR, one or two sensor deployment exhibits poor total, observability except for the 2S1D1P case. The 3S2D1P shows exceptionally strong observability. Target velocity and sound speed are relatively weakly observable when measured by one or two sensors. For small range error, Doppler measurement is crucial with at least two sensors. For small target velocity and sound speed error, 3S2D1P measurement is superior to the other measurement policies.

ACKNOWLEDGEMENT

Research supported by Office of Naval Research Contract No. N000114-81-K-0814.

REFERENCES

1. M. Athans and P.L. Galb, Optimal Control, McGraw-Hill Inc., NY (1966).
2. C.T. Chen, Introduction to Linear System Theory, Holt, Rinehart and Winston, Inc., NY (1970).
3. E.B. Lee and L. Marcus, Foundations of Optimal Control Theory, John Wiley & Sons, NY (1967).
4. M. Hwang and J.H. Seinfeld, "Observability of Nonlinear Systems," J. of Opt. Theory and Appl., 10, 67-77 (1972).
5. R. Hermann and A.J. Krener, "Nonlinear Controllability and Observability," IEEE Tr. Auto. Cont., AC-22, 728-740, (1977).
6. H. Nijeijer, "Observability of Autonomous Discrete Time Nonlinear Systems," Int. J. Control, 36, 867-874 (1982).
7. M. Fliess, "A Remark on Nonlinear Observability," IEEE Tr. Auto. Cont., AC-27, 489-490 (1982).
8. U. Schoenwandt, "On Observability of Nonlinear Systems," 2nd IFAC Itentif. Symposium, Prague Czechoslovakia (1970).
9. Y.M. Kostyukovskii, "Observability of Nonlinear Controlled Systems," Automation Remote Control, 9, 1384-1396 (1968).
10. E.W. Griffith and K.S.P. Kumar, "On the Observability of Nonlinear Systems; I," J. of Math. Anal. and Appl., 35, 135-147 (1971).
11. S.R. Kou, D.L. Elliott, and T.J. Tarn, "Observability of Nonlinear Systems," Info. and Control, 22, 89-99 (1973).

12. T. Fujisawa and E.S. Kuh, "Some Results on Existance and Uniqueness of Sol. of Nonlinear Networks," IEEE Tr. on Circuit th., CT-18, 501-506 (1971).

13. R.S. Palais, "Natural Operation of Differential Forms," Trans. Amer. Math. Soc., 92, 125-141 (1959).

14. R.R. Mohler, C.S. Hwang, "Nonlinear Observability and Mixed Coordinate Bearing-only Signal Analysis," Proc. 23rd Conf. on Decision and Control, Las Vegas (1984).

15. R.S. Liptser and A.N. Shiryayev, Statistics of Random Process II-Applications, Springer-Verlag, NY (1978).

# CHAPTER 6

## LIKELIHOOD RATIOS AND SIGNAL DETECTION FOR NONGAUSSIAN PROCESSES

C.R. Baker and A.F. Gualtierotti

### 1. INTRODUCTION

NonGaussian signal detection problems arise in several applications of underwater acoustics. NonGaussian signal processes occur for active sonar when the reflecting target (with surface undergoing random motions) has only a few dominating scatterers. The noise in such applications is frequently Gaussian, so that the detection problem is that of detecting a nonGaussian signal embedded in additive Gaussian noise.

Problems of detecting a signal in nonGaussian noise also arise; for example, for sonars operating under ice. Noise due to ice-cracking, creaking, floe-smashing, etc., contributes a component which has been found to have substantial nonGaussian behavior [6]. In addition, active sonars operating under ice near the surface may encounter a nonGaussian component due to specular reflection from the irregular under-ice surface. Another environmental situation which may produce nonGaussian noise is shallow-water reverberation.

Optimum detection algorithms require knowledge of the statistical properties of data processes. For applications involving nonGaussian noise with a strong impulsive component, a useful univariate noise model has been developed by Middleton [12]. Much remains to be done in this area. Development of optimum detection algorithms requires knowledge of multivariate statistical models for both the noise and the signal-plus-noise processes. At present, such models do not exist for some of the most important nonGaussian environments. Their development will require a mix of physics, mathematics, statistics, and extensive computational investigations. These are challenging problems whose solution must be obtained before one can obtain optimum detection algorithms.

This contribution first considers algorithms for the detection of nonGaussian signals in Gaussian noise. Results are summarized for the continuous-time problem; more attention is given to discrete-time approximations. A discrete-time recursive algorithm is given. It is shown that (under appropriate assumptions) this discrete-time algorithm is a likelihood ratio detector if the signal-plus-noise process is Gaussian. Attention is then turned to signal detection for problems involving nonGaussian spherically-invariant noise (SIN). The univariate Class A model of Middleton [12] is seen to be a special case of SIN. The likelihood ratio for detection of a signal in SIN is derived for both continuous-time and discrete-time applications. Approximations, including constant false-alarm probability (CFAP) detectors, are discussed. The effect of sample size is also considered. These results indicate that robust detection can be achieved for detection of known signals in spherically-invariant noise. In applying these results to Middleton's Class A model, it is shown that placing that model in the context of SIN provides a number of useful consequences.

All stochastic processes to be discussed are real-valued and defined on a probability space $(\Omega, \beta, P)$. For processes in continuous time, the parameter set is $[0,T]$, and all such processes are assumed to be mean-square continuous. All noise processes are assumed to have zero mean. $(V(t))$ will denote a stochastic process, while $V(t)$ will denote the random variable obtained by sampling the process at the time $t$. The argument $\omega$ in $\Omega$ will typically be suppressed: $V(t) \equiv V(t,\omega)$. $\mathcal{L}_2[0,T]$ is the linear space of all Lebesgue-square-integrable functions on $[0,T]$. $L_2[0,T]$ is the set of equivalence classes $[u]$ obtained from functions $u$ in $\mathcal{L}_2[0,T]$. For a noise process $(N(t))$, $r_N$ will denote the covariance function: $r_N(t,s) = E\, N(t)N(s)$. $R_N$ will denote the covariance operator of $(N(t))$; that is, the integral operator in $L_2[0,T]$ having $r_N$ as its kernel. $R_N$ will be assumed strictly positive; $(\lambda_n)$ is the sequence of (strictly positive) eigenvalues of $R_N$, with $(e_n)$ corresponding c.o.n. eigenvectors. $\langle u,v \rangle = \int_0^T u(t)v(t)dt$ for $u$ and $v$ in $L_2[0,T]$. The signal-plus-noise process at time $t$ will be $Y(t) = S(t) + N(t)$. In the continuous-time case, the likelihood ratio sought is on $L_2[0,T]$: $d\mu_Y/d\mu_N$, where $\mu_Y$ (resp., $\mu_N$) is the probability on $L_2[0,T]$ induced by the stochastic process $(Y(t))$ (resp., $(N(t))$).

For a noise $(N(t))$ with covariance function $r_N$, $H_N$ will denote the reproducing kernel Hilbert space $H_N$ of $r_N$. As is well-known, there is an isometry between $H_N$ and range$(R_N^{\frac{1}{2}})$: the element u is in $H_N$ if and only if there exists a unique element [u] in range$(R_N^{\frac{1}{2}})$ generated by u. The inner product of two elements u and v in $H_N$ is given by $[u,v]_N = \sum_{n\geq 1} \langle [u],e_n\rangle\langle [v],e_n\rangle/\lambda_n$. Since $r_N$ is taken to be continuous, the elements of $H_N$ will be continuous functions. It can also be noted that range$(R_N^{\frac{1}{2}})$ is a real separable Hilbert space under the above inner product (i.e., with respect to the inner product $([u],[v])_N = [u,v]_N$). For simplicity, the element [u] in $L_2[0,T]$ will usually be written simply as u.

For observations $\underline{x}$ in n-dimensional Euclidean space $E^n$, the noise covariance matrix will be assumed strictly positive. $\underline{A}^*$ is the transpose of the matrix $\underline{A}$. For a bounded linear operator A in $L_2[0,T]$, $A^*$ will denote the adjoint.

In discussing existence of likelihood ratios for continuous-time processes, it is necessary to introduce more mathematical structure. Thus, let $(X(t))$, t in $[0,T]$ be a m.s. continuous stochastic process. $\sigma\{X_s, s\leq t\}$ is the $\sigma$-field generated by $\{X_s, s\leq t\}$; $\underline{\sigma}_t^0(X)$ is the filtration consisting of all the $\sigma$-fields $\sigma\{X_s, s\leq t'\}$ for $0 \leq t' \leq t$. $\underline{\sigma}^0(X)$ will denote $\underline{\sigma}_T^0(X)$. $\underline{\sigma}_t(X)$ (resp., $\underline{\sigma}(X)$) will denote the filtration generated by $\underline{\sigma}_t^0(X)$ (resp., $\underline{\sigma}^0(X)$) and all sets of P-measure zero, completed with respect to the underlying probability P. If $(X(t))$ and $(V(t))$ are two such processes, then $\underline{\sigma}_t(V) \vee \underline{\sigma}_t(X)$ will denote the smallest filtration containing both $\underline{\sigma}_t(V)$ and $\underline{\sigma}_t(X)$, with $\underline{\sigma}(V) \vee \underline{\sigma}(X)$ similarly defined.

The continuous-time problems considered here will be modeled in $L_2[0,T]$. Somewhat similar results can be obtained by considering the probabilities induced on $R^{[0,T]}$, the real-valued functions on $[0,T]$. However, those results [5] are not so complete as those for $L_2[0,T]$.

For $\mu_Y$ and $\mu_N$ probabilities on the Borel sets of $L_2[0,T]$, $\mu_Y << \mu_N$ denotes absolute continuity of $\mu_Y$ with respect to $\mu_N$ (so that the likelihood ratio $d\mu_Y/d\mu_N$ exists). $\mu_Y \perp \mu_N$ denotes orthogonal probability measures; in detection applications, orthogonal measures imply singular (perfect) detection. $\mu_Y \sim \mu_N$ denotes mutual absolute continuity: $\mu_Y << \mu_N$ and $\mu_N << \mu_Y$. $P_D$ will denote probability of detection = probability of correctly deciding signal present. $P_{FA}$ denotes probability of false alarm = probability of incorrectly deciding signal present. $P_{FA} = 0$ implies $P_D = 0$ if $\mu_{S+N} << \mu_N$; $P_D = 1$ implies $P_{FA} = 1$ if $\mu_N << \mu_{S+N}$. Thus, $\mu_{S+N} \sim \mu_N$ is the situation usually assumed to hold for practical problems in signal detection. We refer to this as non-singular detection.

Treatment of the continuous-time case introduces substantial complication into the analysis. However, it clarifies structure, enables one to obtain discrete-time finite-sample algorithms as approximations, and provides valuable performance bounds.

## 2. DETECTION OF NONGAUSSIAN SIGNALS IN GAUSSIAN NOISE

An important active sonar application is the detection of nonGaussian signals embedded in additive Gaussian noise. For example, when the noise background is reverberation-limited, the scatterers giving rise to the reverberation can frequently be assumed to be statistically-independent in their reflecting properties. Application of the central limit theorem then gives a Gaussian process for the reverberation process. However, if the target return is primarily due to reflections from a few random scatterers (each contributing random phase and amplitude), then the composite reflection from the target will generally be nonGaussian. In this particular application, the signal and noise processes are dependent, and the noise process is nonstationary.

Other applications may also involve detection of nonGaussian signals in Gaussian noise. For example, in passive sonar the background noise can frequently be assumed to be Gaussian and stationary. However, signal sources such as ship-radiated noise need not be Gaussian.

Full solution of such problems ideally includes determining conditions for nonsingular detection, and then (when nonsingular detection holds) determining the likelihood ratio.

If the signal-plus-noise process is also Gaussian, then conditions for nonsingular detection and the form of the likelihood ratio are well known [4, 14]. If the signal is nonGaussian and independent of the noise, then sufficient conditions for nonsingular detection are given in [3]. With the noise Gaussian, the sufficient condition is that the sample paths of the signal process belong (w.p. 1) to $H_N$, the reproducing kernel Hilbert space of the noise. Under mild assumptions, the likelihood ratio can also be obtained from the results of [3].

If nothing is known about the signal-plus-noise process except its covariance and mean functions, then of course a likelihood ratio detector cannot be determined. However, if one limits consideration to quadratic-linear operations on the data (in forming a test statistic), then the deflection criterion can be used to determine the optimum operation. That is, let $T$ be the class of all admissible test statistics $\tau$. The deflection of $\tau$ is then $D_{01}(\tau) = (E_N\tau(x) - E_{S+N}\tau(x))^2/(E_N\tau^2(x) - [E_N\tau(x)]^2)$, where $E_N(\cdot)$ (resp., $E_{S+N}(\cdot)$) denotes expectation w.r.t. the noise (resp., signal-plus-noise). The problem then consists of determining $\sup_T D_{01}(\tau)$, and determining a $\tau$ achieving this supremum or a sequence $(\tau_n)$ converging to the supremum. The optimum quadratic operation for discrete-time finite-sample data is given in [1], while [2] contains the solution for the infinite-dimensional case and results linking deflection to nonsingular detection.

The results on deflection given in [1] and [2] apply to problems where the signal-plus-noise process is neither Gaussian nor consisting of the noise plus an independent signal process. For such problems, it is desirable to have general conditions for nonsingular detection and also expressions for the likelihood ratio. Currently-available data models may be inadequate to fully utilize such results, but their availability for future use is clearly desirable. Results for the special continuous-time case when the noise is the Wiener process have been known for many years [11]. However, the Wiener process has properties that are not observed in practical sonar problems: sample functions that are almost surely nondifferentiable at almost all time points, the Markov property, and the martingale property. Thus, the design of future optimum signal detection systems requires results beyond those already mentioned; such results have recently been obtained [5].

The results contained in [5] include general conditions for nonsingular detection of a possibly nonGaussian signal imbedded in additive Gaussian noise. The work is based on the spectral representation of second-order stochastic processes, particularly as developed by Hida [10]. The general problem is that of discriminating between a Gaussian noise process $(N(t))$, $t$ in $[0,T]$, and a possibly nonGaussian process $(Y(t))$, $t$ in $[0,T]$.

The basic assumptions made in [5] are the following:

(A.2-1) $(N(t))$ vanishes almost surely at $t = 0$;

(A.2-2) (N(t)) has a purely-continuous spectral representation of multiplicity $M < \infty$.

Assumption (A.2-2) is equivalent to (N(t)) having a representation of the form

$$N(t) = \sum_{i=1}^{M} \int_0^t F_i(t,s)dB_i(s) \qquad (2.1)$$

where $\{(B_i(t)): i \leq M, t \text{ in } [0,T]\}$ is a family of independent-increment mutually-independent path-continuous zero-mean Gaussian processes, and each $F_i$ is a Borel-measurable function on $[0,T]\times[0,T]$ with $F_i(t,s) = 0$ for $s>t$. This representation also satisfies

$\sum_{i=1}^{M} \int_0^T \int_0^T F_i^2(t,s)d\beta_i(s)dt < \infty$, where $\beta_i$ is the Borel measure on $[0,T]$

defined by the non-decreasing variance of $(B_i(t))$:

$\beta_i(a,b] = EB_i^2(b) - EB_i^2(a)$.

The representation (2-1) is taken to be the proper canonical representation for (N(t)) [10]. One consequence is that the completion of the $\sigma$-field $\sigma\{B_i(s): i \leq M, s \leq t\}$ is the same as the completion of $\sigma\{N(s): s \leq t\}$ for each t in [0,T]. In general, the equality (2-1) holds almost surely dP for each fixed t in [0,T]; by assuming that (N(t)) is separable w.r.t. closed sets, one obtains a.s. path equality.

The basic results on non-singular detection of a possibly nonGaussian signal embedded in additive Gaussian noise, as given in [5], entail both a set of sufficient conditions [5, Theorem 2] and a set of necessary conditions [5, Theorem 3]. The sufficient conditions for absolute continuity on $L_2[0,T]$ are given in the following result.

Prop. 2.1 [5, Theorem 3] Let (V(t)) be a stochastic process independent of (N(t)). Suppose that (S(t)) is a stochastic process adapted to $\underline{\sigma}(N) \vee \underline{\sigma}(V)$ and with paths a.s. in $H_N$. If $Y(t) = S(t) + N(t)$ a.e. dtdP, then $\mu_Y << \mu_N$.

Both the sufficient conditions and the necessary conditions include the requirement that the signal process have a representation with almost all paths in the reproducing kernel Hilbert space of the noise covariance function $r_N$. With the representation (2-1), this means that almost all sample functions of the signal process have a representation of the form

$$S(t) = \sum_{i=1}^{M} \int_0^t F_i(t,s)Q_i(s)d\beta_i(s) \qquad (2\text{-}2)$$

where $(Q_i(t))$ is a stochastic process with almost all paths in $\mathcal{L}_2[\beta_i]$:

$\int_0^T Q_i^2(s)d\beta_i(s) < \infty$ a.s. dP. The remaining conditions for absolute continuity embody measurability conditions on the signal process. These conditions are given in terms of the noise process $(N(t))$ and a stochastic process $(V(t))$ independent of the noise. They are essentially related to the signal process being a causal functional of the two processes $(N(t))$ and $(V(t))$. The basic idea is that the signal may be a causal functional of both the noise process (as in the case of dependent signal and noise) and an independent "message" process.

The likelihood ratio $d\mu_{S+N}/d\mu_N$ for this problem is given in [5, Theorem 7]. Define a vector stochastic process $(\underset{\sim}{Z}(t))$ by

$$Z_i(t) = \int_0^t Q_i(s)d\beta_i(s) + B_i(t) \tag{2-3}$$

where the processes $(Q_i(t))$ are those appearing in (2-2) above. Then

$$S(t) + N(t) = \sum_{i=1}^{M} \int_0^t F_i(t,s)dZ_i(s) \tag{2-4}$$

The vector processes $(\underset{\sim}{Z}(t))$ and $(\underset{\sim}{B}(t))$ define probabilities $P_{\underset{\sim}{B}}$ and $P_{\underset{\sim}{Z}}$ on the space of all M-component vector functions on $[0,T]$ whose component functions are all continuous. Under the conditions for existence of $d\mu_{S+N}/d\mu_N$, $dP_{\underset{\sim}{Z}}/dP_{\underset{\sim}{B}}$ will also exist, and for an observation x in $L_2[0,T]$,

$$[d\mu_{S+N}/d\mu_N](x) = [dP_{\underset{\sim}{Z}}/dP_{\underset{\sim}{B}}](\underset{\sim}{m}[x]) \tag{2-5}$$

a.e. $d\mu_N(x)$. $\underset{\sim}{m}(x)$ is an M-component vector of continuous functions, defined by

$$m_i[x](t) = \sum_{n\geq 1} \langle x,e_n\rangle\langle f_t^i,e_n\rangle/\lambda_n, \tag{2-6}$$

$$f_t^i(s) = \int_0^t F_i(s,u)d\beta_i(u).$$

The likelihood ratio $dP_{\underset{\sim}{Z}}/dP_{\underset{\sim}{W}}$ of (2-5) has some explicit known representations [11], depending on the properties of $(\underset{\sim}{Z}(t))$. These representations are based on the fact that each $(B_i(t))$ is a path-continuous Gaussian martingale.

The results given above are for continuous-time observations. In sonar applications, it is desirable to have discrete-time recursive algorithms, which do not require complete recomputation of the test statistic each time a new data point is received. Moreover, it is desirable to have an algorithm with parameters that can be estimated from data, since a complete data model will not usually be available. Such an algorithm will now be derived. It will be based on the following additional assumptions:

(A.2-3) The noise process has multiplicity M=1, and the process $(B_1(t))$ is the standard Wiener process $(W(t))$; thus $N(t) = \int_0^t F(t,s)dW(s)$, where F is a Volterra kernel with $\int_0^T \int_0^T F^2(t,s)dsdt < \infty$;

(A.2-4) The process (Z(t)) defined in (2-3) is a diffusion with respect to the Wiener process and has memoryless drift function, so that $Z(t) = \int_0^t \sigma[Z(s)]ds + W(t)$. (2-7)

The assumption (A.2-3) is reasonable from several viewpoints, such as the fact that multiplicity-one processes are dense (by a mean-square distance criterion) in the space of all second-order processes, and that any Gaussian vector can be represented as the result of a lower-triangular matrix operating on white Gaussian noise. One can also show that the assumption (A.2-3) is satisfied whenever the noise process has a proper canonical representation $N(t) = \int_0^t F(t,s)dB(s)$, where the variance of (B(t)) is an absolutely continuous function on [0,T].

The assumption (A.2-4) is less tenable; it is made primarily for computational convenience (which is in fact not very convenient, even so) when the signal-plus-noise statistics are unknown. It does permit one to consider a very large class of signal-plus-noise processes without having complete knowledge of the statistics. Of course, if a complete mathematical model is available, the assumptions (A.2-3) and (A.2-4) need not be made (if $dP_{\underset{\sim}{Z}}/dP_{\underset{\sim}{B}}$ can be determined).

For the detection problem as defined above, the general form (under a mild restriction) of the likelihood ratio is

$$[d\mu_{S+N}/d\mu_N](x) = \lim_n \exp\,[\Lambda^n(x)]$$

where $0 = t_0^n < t_1^n < t_2^n < \ldots < t_n^n = T$ is a partition of [0,T] such that $\sup_j |t_{j+1}^n - t_j^n| \to 0$,

$$\Lambda^n(x) = \sum_{i=0}^{n-1} \sigma(m[x](t_i^n))(m[x](t_{i+1}^n) - m[x](t_i^n))$$
$$- (1/2) \sum_{i=0}^{n-1} \sigma^2(m[x](t_i^n))(t_{i+1}^n - t_i^n), \tag{2-8}$$

and the limit exists in the norm of $L_1[\mu_N]$.

The representation of $(N(t))$ by $N_t = \int_0^t F(t,s)dW(s)$ yields that $R_N = FF*$, where F is the integral operator with $F(t,s)$ as its kernel, and $F*$ is its adjoint. This can be used to provide an expression for the function m appearing in (2-5) and (2-6) that does not require calculation of eigenvalues and eigenvectors.

First, notice that $\langle e_j, f_t\rangle = \int_0^T \int_0^t F(s,u)du\, e_j(s)ds$ $= \int_0^t \int_0^T F(s,u)e_j(s)dsdu = [LF*e_j](t)$, where $[Lf](t) \equiv \int_0^t f(v)dv$. Using this, the expression (2-6) for m can be rewritten as

$$m[x](t) = \lim_{k\to\infty} [LF* \sum_1^k \langle e_j, x\rangle R_N^{-1} e_j](t)$$
$$= \lim_{k\to\infty} [LF*R_N^{-1}P_k x](t)$$

where $P_k x$ is the projection of the function x on the subspace spanned by $\{e_1, \ldots, e_k\}$. Since $R_N^{-1} = F*^{-1}F^{-1}$, the preceding becomes $m[x](t)$ $= \lim_{k\to\infty} [LF^{-1}P_k x](t)$.

A basic difficulty is that (with probability one [3]) the observation x will not be in the domain of the operator $F^{-1}$, so that $F^{-1}x$ is not defined. In fact, $LF^{-1}$ will in general not be a bounded linear operator. However, for almost all sample functions x (either from noise or signal-plus-noise), $m[x](\cdot)$ is a continuous function of t. Thus the map m is a linear operator from $L_2[0,T]$ into $C[0,T]$ whose domain includes (with probability one) all sample functions of the noise and signal-plus-noise processes.

The difficulty in implementation of the likelihood ratio (2-8) will lie in determining the function $\sigma$ and linear operator m. $\sigma$ is a parameter of the signal-plus-noise process, and its estimation is a problem of considerable interest in stochastic processes (as the drift of a diffusion) and in stochastic filtering. The possibly unbounded linear operator m, mapping $L_2[0,T]$ into $C[0,T]$, depends only on the covariance function of the noise. If the noise covariance function is known, then the preceding expressions can be used to obtain a discrete-time finite-sample approximation to the likelihood ratio. Here we consider such approximations when one knows only the covariance matrix of the noise.

Let $\underline{R}_N$ denote the covariance matrix of the noise; one can write $\underline{R}_N = \underline{F}\ \underline{F}^*$, where the matrix $\underline{F}$ is lower triangular. Now, the expression for m given above is of the form

$$m[x](t) = \lim_{k\to\infty} [LF^{-1}P_k x](t),$$

where $R_N = FF^*$, L is the integration operator, and $P_k$ is the projection of x onto the subspace spanned by $\{e_1,\ldots,e_k\}$, where $\{e_n, n\geq 1\}$ are o.n. eigenvectors of $R_N$. Thus, a reasonable procedure is simply to replace this expression by $m[x] = \underline{L}\ \underline{F}^{-1}\underline{x}$, where $\underline{x}$ is the observed data vector, and $\underline{L}$ is the summation operator in $E^k$; $(\underline{L}\ \underline{x})_j = \sum_{i=1}^{j} x_i$.

There is a fundamental difference between the above approximation to m and the exact result. As previously mentioned, $F^{-1}x$ is (with probability one) not defined for the continuous-time situation; here, of course, there is no such problem for $\underline{F}^{-1}\underline{x}$.

Implementation of the discrete-time algorithm for a fixed sampling interval, $\Delta$, will now be considered. Then, when the observation is an n-component vector, and the above approximation is used, one obtains as an approximation to the log-likelihood ratio the expression

$$\begin{aligned}\Lambda^n(\underline{x}) &= \sum_{j=0}^{n-1} (\sigma[(\underline{L}\ \underline{F}^{-1}\underline{x})_j])[(\underline{L}\ \underline{F}^{-1}\underline{x})_{j+1} - (\underline{L}\ \underline{F}^{-1}\underline{x})_j] \\ &\quad - (\Delta/2) \sum_{j=0}^{n-1} \sigma^2\ [(\underline{L}\ \underline{F}^{-1}\underline{x})_j] \qquad (2\text{-}9) \\ &= \sum_{j=0}^{n-1} (\sigma[(\underline{L}\ \underline{F}^{-1}\underline{x})_j])[(\underline{F}^{-1}\underline{x})_{j+1}] - (\Delta/2) \sum_{j=0}^{n-1} \sigma^2\ [(\underline{L}\ \underline{F}^{-1}\underline{x})_j].\end{aligned}$$

If now a new data point $x_{n+1}$ is observed, the approximation has the recursive form

$$\Lambda^{n+1}(\underline{x}) = \Lambda^n(\underline{x}) + \sigma[(\underline{L}\ \underline{F}^{-1}\underline{x})_n](\underline{F}^{-1}\underline{x})_{n+1} - (\Delta/2)\ \sigma^2[(\underline{L}\ \underline{F}^{-1}\underline{x})_n]. \qquad (2\text{-}10)$$

One notes the following:

(1) Implementation and calculation of $\Lambda$ require the following

operations. First, the function $\sigma$ must be known and programmed. Given the value of $\Lambda^n(\underline{x}^n)$ and the observation $\underline{x}^n = (x_1, \ldots, x_n)$, one stores $\Lambda^n(\underline{x}^n)$, $\underline{x}^n$, $\sigma[(\underline{L}\ \underline{F}^{-1}\underline{x})_n]$, and $(\underline{L}\ \underline{F}^{-1}\underline{x}^n)_n$. When the data point $x_{n+1}$ is received, it is only necessary to use $\underline{x}^{n+1}$ to calculate $(\underline{F}^{-1}\underline{x}^{n+1})_{n+1}$, which means to cross-correlate the observation $\underline{x}^{n+1}$ with the n+1 row of $\underline{F}^{-1}$. This number, say $b_{n+1}$, is then used to form $\Lambda^{n+1}(\underline{x}^{n+1})$,

$$\Lambda^{n+1}(\underline{x}^{n+1}) = \Lambda^n(\underline{x}^n) + \sigma[\sum_1^n b_i]b_{n+1} - (\Delta/2)\ \sigma^2\ [\sum_1^n b_i].$$

This is much simpler than a procedure whereby the function $\underline{m} = \underline{L}\ \underline{F}^{-1}$ is expressed in terms of its eigenvalues and eigenfunctions, since those quantities would have to be stored for $E^n$ and all the sample indices $n \geq 1$, and a complete new calculation done for each new sample point observed.

(2) As already noted, the expression $\Lambda^n(\underline{x}^n)$ can only be considered as an approximation to the discrete-time likelihood ratio. This approximation becomes more valid as n increases, since it amounts to representing the noise vector $\underline{N}$ by $N_i = \sum_{j=1}^{i-1} F_{ij}(\underline{\Delta W})_j$, where $\underline{\Delta W}$ is a vector of i.i.d. $N(0,\Delta)$ random variables. As n increases, $\sum_{j=1}^{i-1} F(i\Delta,\ j\Delta)(\underline{\Delta W})_j$ will converge in mean-square to $N(i\Delta) = N(t)$, keeping $i\Delta = t$, where the function F is that appearing in the representation $N_t = \int_0^t F(t,s)dW_s$. Thus, as n increases, the representation of $\underline{N}$ by $N(t) = \sum_{j=1}^{i-1} F_{ij}(\underline{\Delta W})_j$ converges to the representation satisfying the original continuous-time models for noise and signal-plus-noise.

(3) $\sigma$ can be estimated from a sample of data representative of signal-plus-noise. In discrete time, the procedure is as follows, given an observed $\underline{S}+\underline{N}$ vector $\underline{X}$.

a) Form $\underline{\Delta Z} = \underline{F}^{-1}\underline{X}$, where $\underline{R}_N = \underline{F}\ \underline{F}^*$, F lower-triangular, $(\underline{\Delta Z})_i = Z(i\Delta) - Z([i-1]\Delta)$, $Z_0 = 0$.

b) $Z(i\Delta) = \Delta \sum_{j=1}^{i-1} \sigma(Z_{j\Delta}) + W(i\Delta)$.

Given the sample vector $\underline{Z}$ obtained from b), the function $\sigma$ can be estimated. A maximum-likelihood procedure is given in [8].

Of course, the approximations (2-8) and (2-9) need not be likelihood ratios for a fixed finite set of sample points. However, it will now be shown that (2-9) is a likelihood ratio when the function $\sigma$ is linear. In this case, $\underline{S}+\underline{N}$ is Gaussian, so that the likelihood ratio $d\mu_{\underline{S}+\underline{N}}/d\mu_{\underline{N}}$ can be found.

In accord with the model for $(Z(t))$, the discrete-time representation is (for $\sigma$ linear)

$$Z_{k+1} = \Delta \sum_{j=1}^{k} a_j Z_j + W_{k+1}. \tag{2-11}$$

It will be shown that for an observation vector $\underline{x}$ in $E^n$,

$$-\underline{x}*(\underline{R}_Z^{-1} - \underline{R}_W^{-1})\underline{x}/2 = -(\Delta/2) \sum_{i=1}^{n-1} a_i^2 x_i^2 + \sum_{i=1}^{n-1} a_i x_i (x_{i+1} - x_i) \tag{2-12}$$

The LHS of (2-12) is the log-likelihood ratio (within a constant) of $dP_{\underline{Z}}/dP_{\underline{W}}$. Given the equality (2-12), if one has that $\underline{N} = \underline{F}\ \underline{\Delta W}$, $\underline{S}+\underline{N} = F\ \underline{\Delta Z}$, then $[d\mu_{\underline{S}+\underline{N}}/d\mu_{\underline{N}}](\underline{x}) = [dP_{\underline{\Delta Z}}/dP_{\underline{\Delta W}}](\underline{F}^{-1}\underline{x})$.

Thus let

$$Z_{k+1} = \Delta \sum_{j=1}^{k} a_j Z_j + W_{k+1}, \qquad k \geq 1,$$

$$Z_1 = W_1.$$

Let $\underline{A}$ be the matrix $\text{diag}[a_1, \ldots, a_n]$. The RHS of (2-9), evaluated at $\underline{y} = \underline{F}\ \underline{L}^{-1}\underline{x}$, then becomes

$$\left[ \sum_{j=1}^{n-1} (Ax)_j (x_{j+1} - x_j) - \tfrac{1}{2} \Delta \sum_{j=1}^{n-1} (Ax)_j^2 \right]. \tag{2-13}$$

To show that (2-9) is a likelihood ratio test statistic, it will first be shown that (2-13) is equal to $-\underline{x}*(\underline{R}_Z^{-1} - \underline{R}_W^{-1})\underline{x}/2 = \log [dP_{\underline{Z}}/dP_{\underline{W}}](\underline{x})$ + constant.

The above representation for $\underline{Z}$ gives

$$(\underline{I} + \Delta \underline{A})\underline{Z} = \Delta \underline{L}\ \underline{A}\ \underline{Z} + \underline{W}$$

so

$$\underline{Z} = \underline{B}^{-1}\underline{W}$$

$$\underline{B} = I + \Delta \underline{A} - \Delta \underline{L}\ \underline{A}.$$

$\underline{Z}$ thus has covariance matrix $\underline{R}_Z = \underline{B}^{-1}\underline{R}_W \underline{B}*^{-1}$, so $\underline{R}_Z^{-1} = \underline{B}*\underline{R}_W^{-1}\underline{B}$. Since $\underline{R}_W(i,j) = \Delta \min(i,j)$, $\underline{R}_W = \Delta LL*$, and thus

$$\underline{R}_Z^{-1} = (I + \Delta\underline{A} - \Delta\underline{A}\ \underline{L}*)\underline{L}*^{-1}\underline{L}^{-1}(\underline{I} + \Delta\underline{A} - \Delta\underline{L}\ \underline{A})/\Delta$$

where $(\underline{L}^{-1})_{ij} = 1$ if $i=j$

$= -1$ if $i=j+1$

$= 0$ otherwise.

This gives $\underline{R}_Z^{-1} - \underline{R}_W^{-1}$

$$= [\Delta\underline{A}\underline{L}*^{-1}\underline{L}^{-1}\underline{A} + \underline{A}\underline{L}*^{-1}\underline{L}^{-1} + \underline{L}*^{-1}\underline{L}^{-1}\underline{A} - \underline{A}\underline{L}^{-1}(\underline{I}+\Delta\underline{A}) - (\underline{I}+\Delta\underline{A})\underline{L}*^{-1}\underline{A} + \Delta\underline{A}^2]$$

and for a data vector $\underline{x}$,

$$\underline{x}*(\underline{R}_Z^{-1} - \underline{R}_W^{-1})\underline{x} = \Delta(\underline{L}^{-1}\underline{Ax})*\underline{L}^{-1}\underline{Ax} + 2(\underline{L}^{-1}\underline{x})*\underline{L}^{-1}\underline{Ax}$$
$$- 2(\underline{L}^{-1}\underline{x})*\underline{Ax} - 2\Delta(\underline{L}^{-1}\underline{Ax})*\underline{Ax} + \Delta(\underline{Ax})*\underline{Ax}.$$

The three terms containing $\Delta$ sum to $\Delta \sum_{i=1}^{n-1} a_i^2 x_i^2$, while the other two terms sum to $-2 \sum_{i=1}^{n-1} a_i x_i (x_{i+1} - x_i)$, so that

$$\underline{x}*(\underline{R}_Z^{-1} - \underline{R}_W^{-1})\underline{x} = -2 \sum_{j=1}^{n-1} a_j x_j (x_{j+1} - x_j) + \Delta \sum_{j=1}^{n-1} a_j^2 x_j^2, \text{ as desired.}$$

This shows that (2-9), evaluated at $\underline{y} = \underline{F}\ \underline{L}^{-1}\underline{x}$, satisfies $\Lambda^n(\underline{F}\ \underline{L}^{-1}\underline{x}) = -\underline{x}*(\underline{R}_Z^{-1} - \underline{R}_W^{-1})\underline{x}/2 = \log [dP_{\underline{Z}}/dP_{\underline{W}}](\underline{x}) + \text{constant}$. Moreover, $[d\mu_{\underline{S+N}}/d\mu_{\underline{N}}](\underline{y}) = [dP_{\underline{\Delta Z}}/dP_{\underline{\Delta W}}](\underline{F}^{-1}\underline{y}) = [dP_{\underline{Z}}/dP_{\underline{W}}](\underline{L}\ \underline{F}^{-1}\underline{y})$, the last equality because $\underline{\Delta Z} = \underline{L}^{-1}\underline{Z}$, $\underline{\Delta W} = \underline{L}^{-1}\underline{W}$. With $\underline{y} = \underline{F}\ \underline{L}^{-1}\underline{x}$, $[d\mu_{\underline{S+N}}/d\mu_{\underline{N}}](\underline{y}) = [dP_{\underline{Z}}/dP_{\underline{W}}](\underline{x}) =$ (from above) $\exp[\Lambda^n(\underline{y}) + \text{constant}]$.

Thus, when the above assumptions are satisfied (including the assumption that $\underline{Z}$ is a Gaussian vector), the approximation given in (2-9) is a discrete-time finite-sample likelihood ratio.

## 3. SPHERICALLY-INVARIANT NOISE (SIN) MODELS

Let $(N(t))$, $t$ in $T$, be a real-valued zero-mean stochastic process on a probability space $(\Omega, \beta, P)$. N is said to be spherically invariant if it has the representation $N(t) = AG(t)$ for each $t$ in $T$, where G is a Gaussian process and A is a random variable which is independent of G and which has finite second moment. Since it can be assumed that $EA^2 = 1$, the covariance of N can be taken to be the same as that of G. Thus, the finite-dimensional distributions of N are completely determined by its covariance and by the distribution of the random variable A. SIN can thus be viewed as a first step away from Gaussian noise.

If the random variable A is discrete, then the distribution of the random vector $(N(t_1),\ldots,N(t_r))$ is given by the density function

$$f(\underline{N}) = \sum_{i=1}^{K} p_i n(0,a_i^2 R) \tag{3-1}$$

where n(a,B) is the density of a Gaussian random vector (in $E^r$) with mean a and covariance matrix B. In the representation (3-1), $P[A=a_i] = p_i$, and $K \leq \infty$ is the number of distinct values that A assumes with positive probability. In this paper, it will be assumed throughout that A is a discrete random variable. We also assume (WLOG) that $EA^2 = 1$ and that A is strictly positive.

The model for univariate impulsive-plus-Gaussian noise developed by Middleton [12] takes two basic forms, defined as Class A and Class B, depending on the relative bandwidth of noise and receiver. The Class A model is defined to exist when the impulsive noise pulses do not cause transients in the front end of the receiver; it is thus a model for narrowband noise. The univariate density function as developed by Middleton has the form [12]

$$f(x) = e^{-U} \sum_{m=0}^{\infty} \frac{U^m}{m!} \frac{1}{\sqrt{2\pi}\, a_m} \exp\left\{-\tfrac{1}{2} x^2/a_m^2\right\} \tag{3-2}$$

where U is the "overlap index" and $(a_m^2)$ is a sequence of variance components. The overlap index is defined to be the average number of arrivals per second multiplied by the average length of the pulse. The variance component $a_m^2$ is defined by $a_m^2 = (mU^{-1} + \Gamma)/(1 + \Gamma)$, where $\Gamma$ is the ratio of the intensities of the Gaussian and non-Gaussian components of the noise.

It can be seen that (3-2) is the probability distribution of a spherically-invariant random variable X = AY, where Y is a zero-mean unit-variance Gaussian r.v., and A is an independent r.v. taking the values $(a_m)$ with

$$p_m = P[A=a_m] = U^m e^{-U}/m! \tag{3-3}$$

In fact, $E\,A^2 = \sum_{m=0}^{\infty} \frac{U^m e^{-U}}{m!} (\frac{m}{U} + \Gamma)/(1+\Gamma) = 1.$

In [16], Spaulding and Middleton analyse the problem of detecting a known signal in Class A noise by assuming independent sampling, so that the sampled noise data has joint density function (n samples)

$$p(\underline{x}) = \prod_{i=1}^{n} \sum_{m=0}^{\infty} p_m \frac{1}{\sqrt{2\pi}\, a_m} \exp\left[-\frac{x_i^2}{2a_m^2}\right]. \tag{3-4}$$

However, if the r.v. A is constant over the sampling interval, then the density of $\underline{X} = A\underline{Y}$, where the components of $\underline{Y}$ are i.i.d. N(0,1), is

$$p(\underline{x}) = \sum_{m=0}^{\infty} p_m \frac{1}{\left[2\pi\, a_m^2\right]^{n/2}} \exp\left[-\frac{\|x\|^2}{2a_m^2}\right] \tag{3-5}$$

where $\|x\|^2 \equiv \sum_{i=1}^{n} x_i^2$. When the Gaussian process Y has non-singular covariance matrix R, then the class A noise has joint density (if the r.v. A is constant over the observation interval)

$$p(\underline{x}) = \sum_{m=0}^{\infty} p_m \frac{1}{\left[2\pi\, a_m^2\right]^{n/2} (\det R)^{\frac{1}{2}}} \exp\left[-\tfrac{1}{2}\, \underline{x}^* R^{-1} \underline{x}/a_m^2\right] \tag{3-6}$$

As will be shown in the next section, for reasonably large n it is not necessary to know the values of U and $\Gamma$ in order to implement this detector. This fact, as well as the joint density (3-6), illustrates some of the advantages of using a general SIN model whenever appropriate.

Of course, SIN models are not limited to the Middleton model. They cover a large family of smooth unimodal densities that are symmetric about their mean. NonGaussian examples of spherically-invariant distributions include the t and double-exponential [9].

## 4. DETECTION IN SPHERICALLY-INVARIANT NOISE (SIN)

In this section, (N(t)) will be SIN with representation (AG(t)). (G(t)) is a m.s. continuous zero-mean Gaussian process and A is a strictly-positive discrete random variable independent of (G(t)) and with $EA^2 = 1$. (N(t)) thus has zero mean and covariance the same as the covariance of (G(t)). A takes on the value $a_i$ with probability $p_i > 0$.

Likelihood ratio detection of a known signal in SIN has previously been considered by Yao [18] for a very special case: the threshold on the likelihood ratio test statistic is unity. It has also been considered by Spooner [17] for a specific distribution of the mixing random variable A. A more comprehensive treatment has been given by Picinbono and Vezzosi [13]. Their work, and that of Spooner and Yao, has been for the discrete-time finite-sample-size problem.

However, these authors all use or permit continuous mixing r.v.'s A. Our choice of a discrete r.v. for A permits analysis of the continuous time problem without introducing much mathematical complication. It is also sufficient to apply our results to detection in Middleton's Class A noise.

The first topic to be addressed here is that of absolute continuity and likelihood ratio. Sufficient conditions are contained in the following result.

Prop. 4.1. Suppose that $(Y(t))$ is a stochastic process adapted to $\underline{\sigma}(G) \vee \underline{\sigma}(V)$, where $(V(t))$ is any process independent of $(N(t))$. Suppose also that $Y(t) = S(t) + N(t)$ a.e. $dtdP$, where $(S(t))$ is a stochastic process adapted to $\underline{\sigma}(Y)$ and with almost all paths in $H_N$. Then $\mu_Y \ll \mu_N$. Moreover, $\mu_{S+aG} \ll \mu_{aG}$ for all $a>0$, and

$$[d\mu_{S+N}/d\mu_N](x) = \Sigma_i \, I_{C(a_i)}(x) \, [d\mu_{S+a_iG}/d\mu_{a_iG}](x) \qquad (4\text{-}1)$$

a.e. $d\mu_N(x)$. In (4-1), the sum is over all $a_i$ such that $P[A=a_i] > 0$, $I_C$ is the indicator function for the set C in $L_2[0,T]$, and

$$C(a_i) = \{x\colon \lim_n \frac{1}{n} \sum_{j=1}^{n} \langle x, e_j\rangle^2/\lambda_j = a_i^2\}.$$

Moreover, if $(S(t))$ is any process such that $P[A=a_i] > 0$ implies $\mu_{S+a_iG} \ll \mu_{a_iG}$, then $\mu_{S+N} \ll \mu_N$ and $d\mu_{S+N}/d\mu_N$ has the representation (4-1).

Proof. If $(Y_t)$ is adapted to $\underline{\sigma}(G) \vee \underline{\sigma}(V)$, then $(Y_t)$ is adapted to $\underline{\sigma}(aG) \vee \underline{\sigma}(V)$ for any constant a. Since $H_N = H_G = H_{aG}$, $\mu_{S+aG} \ll \mu_{aG}$ for any positive constant a, by Theorem 3 of [5]. Then for any Borel set B of $L_2[0,1]$,

$$\mu_{S+N}(B) = \Sigma_i \, p_i \mu_{S+a_iG}(B) = \Sigma_i \, p_i \int_B [d\mu_{S+a_iG}/d\mu_{a_iG}](x) \, d\mu_{a_iG}(x).$$

Now, $\lim_{n\to\infty} \frac{1}{n} \sum_{j=1}^{n} \langle x, e_j\rangle^2/\lambda_j = a_i^2$ w.p.1 when $A = a_i$, under both $\mu_N$ and $\mu_{S+N}$. To see this for $\mu_N$ (noise-only data), one notes that the random variable $\langle x, e_j\rangle/\lambda_j^{\frac{1}{2}}$ has the form $a_i\langle G, e_j\rangle/\lambda_j^{\frac{1}{2}}$ when only noise is present and $A = a_i$. The random variables $\{\langle G, e_j\rangle/\lambda_j^{\frac{1}{2}}\colon j \geq 1\}$ are i.i.d. $N(0,1)$. Thus, by the law of large numbers, $\frac{1}{n} \sum_{1}^{n} \langle G, e_j\rangle^2/\lambda_j \to 1$ with probability one. When $A=a_i$, $\frac{1}{n} \sum_{j=1}^{n} \langle N, e_j\rangle^2/\lambda_j \to a_i^2$ w.p.1 $(\mu_N)$. If signal is present and $A=a_i$, then $\lim_{n\to\infty} \frac{1}{n} \sum_1^n \langle S+a_iG, \lambda_j\rangle^2/\lambda_j$ is again equal to $a_i^2$ w.p.1. To see this, note that since S is in $H_N$ w.p. 1, $\sum_1^\infty \langle S, e_j\rangle^2/\lambda_j$ is finite. This implies that both $\frac{1}{n} \sum_1^n \langle S, e_j\rangle^2/\lambda_j$ and $\frac{1}{n} \sum_1^n \langle S, e_j\rangle\langle a_iG, e_j\rangle/\lambda_j$ converge to zero w.p. 1.

The preceding shows that $\mu_{a_iG}[C(a_i)] = 1$ and that $\mu_{a_jG}[C(a_i)] = 0$ for $i\neq j$. Thus,

$$\mu_{S+N}(B) = \sum_i p_i \int_{B\cap C(a_i)} [d\mu_{S+a_iG}/d\mu_{a_iG}](x)\, d\mu_{a_iG}(x)$$

$$= \sum_i \int_{B\cap C(a_i)} [d\mu_{S+a_iG}/d\mu_{a_iG}](x)\, d\mu_N(x)$$

because $\mu_N(B\cap C(a_i)) = p_i\mu_{a_iG}(B\cap C(a_i))$. (4-1) now follows by the monotone convergence theorem.

□

The expression for the likelihood ratio given in (4-1) partitions the Borel $\sigma$-field of $L_2[0,T]$ into two major subsets. These sets are $\bigcup_i C(a_i)$ and its complement, where $C(a_i) = \{x: \lim_n \frac{1}{n} \sum_{j=1}^n \langle x, e_j\rangle^2/\lambda_j = a_i^2\}$. It is noteworthy that the likelihood ratio does not involve the probabilities $P\{A=a_i\}$. These facts show first that the important factor in determining the likelihood ratio for detection in continuous-time SIN is knowledge of the values which can be assumed by the mixing random variable A. However, no penalty is assessed if one includes too many possible values of A. That is, if b is not a possible value of A, then the set $C(b) = \{x: \lim_n \frac{1}{n} \sum_{j=1}^n \langle x, e_j\rangle^2/\lambda_j = b^2\}$ has zero $\mu_N$-probability, and so addition of the term $I_{C(b)}(x)[d\mu_{S+bG}/d\mu_{bG}](x)$ to (4-1) will not affect (with probability one) performance of the test statistic.

A particular application of the above is the situation when the noise can be either Gaussian or spherically-invariant nonGaussian. If $P[A=1] > 0$ holds for the mixing r.v. in the nonGaussian case, then the likelihood ratio (4-1) will still be a likelihood ratio if the noise is in fact Gaussian. If $P[A=1] = 0$ for the nonGaussian SIN model, then one can add the term $I_{C(1)}(x) \dfrac{d\mu_{S+G}}{d\mu_G}(x)$ to the likelihood ratio (4-1).

The resulting sum will be a likelihood ratio when either hypothesis is true.

In the remainder of this section, attention will be restricted to the problem of detecting a known signal S in additive SIN. In the case of Gaussian noise, it is well known that the likelihood ratio exists (non-singular detection) if and only if S is in range($R_N^{\frac{1}{2}}$). The same result holds if the noise is any SIN process.

<u>Prop. 4.2</u>. If S is a fixed element in $\mathscr{L}_2[0,T]$, then either $\mu_{S+N} \perp \mu_N$ or else $\mu_{S+N} << \mu_N$ and $\mu_N << \mu_{S+N}$. Mutual absolute continuity holds if and only if S is in range($R_N^{\frac{1}{2}}$).

<u>Proof</u>. If the $L_2[0,T]$ equivalence class generated by S is not in range($R_N^{\frac{1}{2}}$), then $\mu_{S+a_iG} \perp \mu_{a_iG}$ for each $a_i$, using the known results for the Gaussian case. As shown in the proof of Prop. 4.1 (and well-known [14]), $\mu_{a_iG} \perp \mu_{a_jG}$ for $i \neq j$. Moreover, $\mu_{S+a_iG} \perp \mu_{a_jG}$ for $j \neq i$, since range($R_N^{\frac{1}{2}}$) = range$\left[(a_i^2 R_N + a_j^2 R_N)^{\frac{1}{2}}\right]$ and S must belong to this latter range space in order to have mutual absolute continuity of $\mu_{S+a_iG}$ and $\mu_{a_jG}$ [14]. Thus $\mu_{S+a_iG} \perp \mu_{a_jG}$ for all i and j, so that $\mu_{S+a_iG} \perp \sum_{j=1}^{K} p_j \mu_{a_jG}$ for $i=1,\ldots,K$. This gives $\sum_{i=1}^{K} p_i \mu_{S+a_iG} \perp \sum_{j=1}^{K} p_j \mu_{a_jG}$, or $\mu_{S+N} \perp \mu_N$.

Conversely, if S is in range($R_N^{\frac{1}{2}}$), $\mu_{S+a_iG} \sim \mu_{a_iG}$ for $i=1,\ldots,K$, so that

$$\sum_{i=1}^{K} p_i \mu_{S+a_iG} \sim \sum_{j=1}^{K} p_j \mu_{a_jG}. \qquad \square$$

Performance of the likelihood ratio (4-1) can be computed for the case of a known signal. When the noise is Gaussian, then it is well-known that the performance depends only on

$$d^2 = \|S\|_N^2 = \sum_{n\geq 1} \langle S, e_n \rangle^2 / \lambda_n$$

where $\{\lambda_n, n\geq 1\}$ and $\{e_n, n\geq 1\}$ are the eigenvalues and associated c.o.n. eigenvectors of the noise covariance operator $R_N$.

Prop. 4-3. Suppose that the signal is a known function S belonging to range($R_N^{\frac{1}{2}}$), and that $P[A=a_i] = p_i > 0$, $i \geq 1$. Then performance of a likelihood ratio test statistic is given by

$$P_{FA} = \sum_i p_i \; P[Z \geq ka_i + d/(2a_i)] \qquad (4\text{-}2)$$

$$P_D = \sum_i p_i \; P[Z \geq ka_i - d/(2a_i)] \qquad (4\text{-}3)$$

where Z is distributed N(0,1) and k is a constant whose value is determined by the desired value of $P_{FA}$.

Proof: For $k > 0$,

$$P_{FA} = \mu_N\{x: (d\mu_{S+N}/d\mu_N)(x) \geq e^{kd}\} = \mu_N\{x: \sum_i I_{C(a_i)}(x)\ell_i(x) \geq kd\}$$

where

$$\ell_i(x) = \log[(d\mu_{S+a_iG}/d\mu_{a_iG})(x)] = \frac{1}{a_i^2}\left[\sum_n \langle x,e_n\rangle\langle S,e_n\rangle/\lambda_n - d^2/2\right].$$

Thus

$$P_{FA} = \sum_i p_i \; \mu_{a_iG}\{x: \ell_i(x) \geq kd\}$$

$$= \sum_i p_i\mu_G\{x: a_i\Sigma_n \langle x,e_n\rangle\langle S,e_n\rangle/\lambda_n \geq kda_i^2 + d^2/2\}.$$

Since the random variable $\ell$ defined by

$$\ell(x) = \Sigma_n \langle x,e_n\rangle\langle S,e_n\rangle/\lambda_n$$

is Gaussian with respect to $\mu_G$, and has mean zero and variance $d^2$,

$$P_{FA} = \Sigma_i \; p_i \; P[Z \geq ka_i + d/(2a_i)].$$

$P_D$ is calculated in the same way.

□

As can be seen from (4-2) and (4-3), detection performance depends on d and also on the distribution of the mixing random variable A.

The likelihood ratio as given in (4-1) requires prior knowledge of the values $(a_i)$ that can be assumed by the mixing random variable A. However, this prior knowledge is not necessary in order to implement this detector.

Prop. 4-4. For a known signal S, a likelihood ratio test is to decide signal present if and only if

$$[\ell(x)/\hat{A}^2(x) - d^2/(2\hat{A}^2(x)] \geq k \qquad (4\text{-}4)$$

where k is determined from (4-2) and (4-3), and

$$\hat{A}^2(x) = \lim_n \frac{1}{n} \sum_{k=1}^{n} \langle x, e_k \rangle^2 / \lambda_k .$$

<u>Proof</u>. If $x \in C(a_i)$, then $\hat{A}^2(x) = a_i^2$ with probability one under $\mu_N$ or $\mu_{S+N}$, as shown in the proof of Prop. 4-1. The result then follows directly from the expression (4-1), or by examining the proof of Prop. 4-3.

□

A likelihood ratio detector can thus be implemented without any prior knowledge of the distribution of the mixing random variable A, provided the noise is in fact SIN. However, as can be seen from the expressions (4-2) and (4-3) for $P_{FA}$ and $P_D$, likelihood ratio detection performance depends on the complete distribution of A. This means that it is not possible to set a threshold for a specified $P_{FA}$ unless one has complete knowledge of the distribution of A.

This leads one to consider the problem of CFAP (constant false alarm probability) detection, which has been treated for many years by designers of active sonar detection systems. In this traditional context, the detection problem is that of detecting a signal in Gaussian noise which is known except for a scale factor. It is desired to have the same probability of false alarm for any value of the scale factor. The scale factor has usually been treated as an unknown parameter, rather than as a random variable.

A CFAP detector can be obtained for the SIN detection problem by using the following decision procedure:

decide signal present if and only if

$$\ell(x)/\hat{A}(x) \geq kd, \tag{4-5}$$

when $\hat{A}(x) = [\hat{A}^2(x)]^{\frac{1}{2}}$. When noise only is present, $\ell(x)/\hat{A}(x) = \ell(x)/a_i$ with probability one when $x = a_i G$. Since then $\ell(x) = a_i \ell(G)$, one has that $\ell(x)/\hat{A}(x)$ is Gaussian with zero mean and variance $d^2$ and

$$P_{FA} = P[Z \geq k]. \qquad (4\text{-}6).$$

When this detection algorithm is used, and $x = S + a_i G$, then $\ell(x)/a_i$ is Gaussian with mean $d^2/a_i$ and variance $d^2$, so that

$$P_D = \Sigma\, p_i\, P[Z \geq k - d/a_i]. \tag{4-7}$$

The difference in performance between the optimum detector (4-4) and the CFAP detector (4-5) will depend on the distribution of A. Figure 1 shows an example using a distribution for A obtained from analyzing under-ice sonar data. The curves show performance for the optimum detector (4-4), the CFAP detector (4-5), and the matched filter (A=1 w.p. 1) when the noise is SIN with the given distribution for A. Also shown is the performance that one would obtain using the matched filter if the noise were truly Gaussian. The difference in performance of the matched filter and the likelihood ratio illustrates the significant performance loss that can occur if the noise is mistakenly assumed to be Gaussian. This and the similarity in performance of the CFAP detector and the likelihood ratio illustrate the wisdom of using a CFAP detector if there is a possibility that the noise is SIN with unknown distribution.

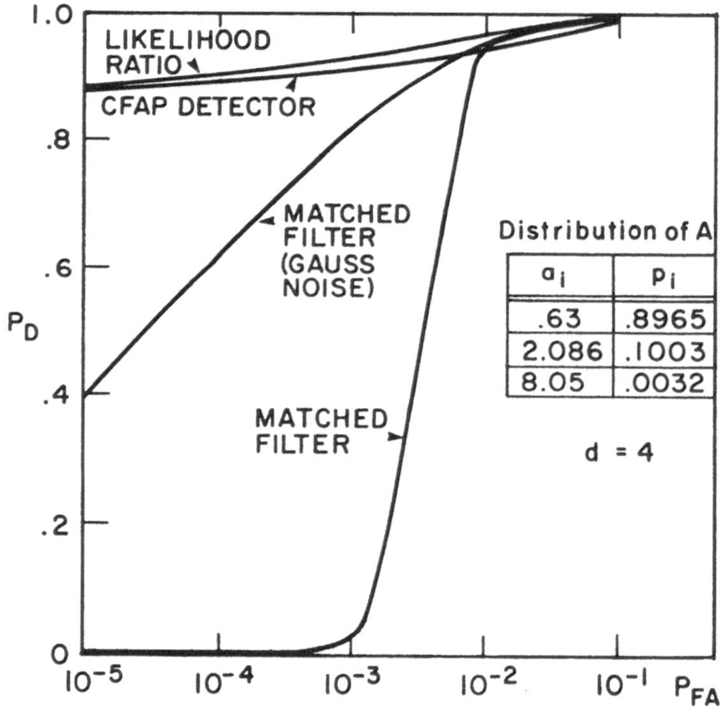


Figure 1. Detection of Known Signal in SIN

For the discrete-time finite-sample detection problem, with observation $\underline{x}$ in $E^n$, the likelihood ratio $dP^n_{\underline{S}+\underline{N}}/dP^n_{\underline{N}}$ is easily seen to be

$$\left[dP^n_{\underline{S}+\underline{N}}/dP^n_{\underline{N}}\right](\underline{x}) = \sum_{i=1}^{K} p_i dP^n_{\underline{S}+a_i\underline{G}}(\underline{x}) / \sum_{j=1}^{K} p_j dP^n_{a_j\underline{G}}(\underline{x}) \qquad (4\text{-}8)$$

where $dP^n$ is the multivariate density function for the probability $P^n$ on $E^n$. In contrast to the continuous-time case (4-1), the probabilities $p_i = P[A=a_i]$ appear in (4-8). Moreover, the likelihood ratio (4-1) produces a non-zero value (with probability one) only if the observation involves $a_iG$ for $a_i$ one of the terms included in (4-1); otherwise, the value of the likelihood ratio is zero. This is not true in the discrete-time case of (4-8). In the case of a known signal S, the performance ($P_{FA}$ and $P_D$) of (4-1) depends only on the distribution of A and on $d^2 = \|R_N^{-\frac{1}{2}}S\|^2$. The discrete-time detector's performance improves as the sample size n increases, with d fixed.

These differences can all be understood by examining the form of (4-8) as the sample size increases.

Suppose that noise only is present, and that the mixing r.v. A takes on the value $a_i$, so that the received waveform $\underline{x}$ is $a_i\underline{G}$ with $\underline{G}$ multivariate Gaussian, zero mean, non-singular covariance matrix $\underline{R}$.

Let $Y^n_j = \log p_j \, dP^n_{a_j\underline{G}}(\underline{x}) = \log p_j - n \log a_j - a_i^2\|\underline{R}^{-\frac{1}{2}}\underline{G}\|^2/(2a_j^2) + C$

$$= \log p_j - n \log a_j - a_i^2 \sum_{k=1}^{n} V_k^2/(2a_j^2) + C$$

where C is a constant and $(V_k)$ is i.i.d. N(0,1). It will be shown that $Y^n_j - Y^n_i \to -\infty$ w.p. 1. Thus,

$$Y^n_j - Y^n_i = \log [p_j/p_i] - n \log [a_j/a_i] + \sum_{k=1}^{n} V_k^2\left[\tfrac{1}{2}(1 - a_i^2/a_j^2)\right]$$

$$= \gamma - n\rho u_{i,j} + \beta\chi_n^2 u_{i,j}$$

where $\chi_n^2$ is chi-square with n degrees of freedom, $\beta = \beta(i,j) = |(1-a_i^2/a_j^2)/2|$, $\rho = \rho(i,j) = |\log (a_i/a_j)|$, $\gamma = \gamma(i,j) = \log p_j/p_i$, and $u_{i,j} = 1$ if $a_i \le a_j$, $u_{i,j} = -1$ if $a_i > a_j$.

For $a_i < a_j$, $m > 0$, $\underline{x} = a_i\underline{G}$, one has $P[Y^n_j - Y^n_i \ge -m]$ $= P[p_j dP_{a_jG}(\underline{x}) \ge e^{-m} p_i dP^n_{a_i\underline{G}}(\underline{x})] = P[\chi_n^2 \ge -(m+\gamma)/\beta + n\rho/\beta]$.

Since $\chi_n^2/n \to 1$ w.p. 1 as $n \to \infty$, and $\rho/\beta > 1$ for $a_i < a_j$,

$P[\lim_n (\chi_n^2/n + (m+\gamma)/\beta n - \rho/\beta) \geq 0] = 0$, for any fixed $m > 0$.

If $a_i > a_j$, then $P[Y_j^n - Y_i^n \geq -m] = P[\chi_n^2 \leq (m+\gamma)/\beta + n\rho/\beta]$. In this case, $\rho/\beta < 1$, so $P[\lim_n(\chi_n^2/n - (m+\gamma)/\beta n - \rho/\beta) \leq 0] = 0$.

Using these expressions, one can obtain the value of n required for a specified approximation, once the distribution of A is known. The value of n required for $P[p_j dP_{a_j \underline{G}}(\underline{x}) \geq e^{-m} p_i dP_{a_i \underline{G}}(\underline{x})] \leq \alpha$ when $\underline{x} = a_i \underline{G}$ is determined from

$$P[\chi_n^2 \geq -(m+\gamma)/\beta + n\rho/\beta] \leq \alpha \qquad \text{if } a_i < a_j$$
$$P[\chi_n^2 \leq (m+\gamma)/\beta + n\rho/\beta] \leq \alpha \qquad \text{if } a_i > a_j. \tag{4-9}$$

This procedure can be repeated for the numerator of (4-8). A conservative result, which satisfies $P[p_j dP_{\underline{S}+a_j \underline{G}}(\underline{x}) \geq e^{-m} p_i dP_{\underline{S}+a_i \underline{G}}(\underline{x})] \leq \alpha$, is to require $n \geq \Delta n + n_1$, where $n_1$ is the value for $\alpha$ given by (4-9) and $\Delta n$ satisfies

$$\Delta n \rho(i,j)/\beta(i,j) \geq dZ_\alpha/a_i + d^2 u(i,j)/(2a_i^2). \tag{4-10}$$

Applying these results to the distribution used to obtain Figure 1, with $e^{-m} = .01$ ($m = 4.6$) and $\alpha = 10^{-3}$, the required sample size for the denominator of (4-8) is $n \geq 21$, using (4-9). The value of $\Delta n$ given by (4-10) is 9, so that an adequate sample size is 30. This rather small required sample size is a result of the wide separation between the three values of A, and (to a much lesser extent) the corresponding large differences in their probabilities. It can be seen from (4-9) and (4-10) that distributions for A which have more similar values will require larger sample sizes in order to achieve the above bounds, with a requirement of $n \to \infty$ as the minimum distance between A values converges to zero.

The gist of this analysis is that the likelihood ratio (4-8) converges to $dP_{\underline{S}+a_i \underline{G}}/dP_{a_i \underline{G}}$ when $\underline{x} = a_i \underline{G}$, as the sample size increases; the rate of convergence depends on the distance from $a_i$ to the nearest value of A not equal to $a_i$; and the probability ratios can be ignored for large sample size. When the sample size n is sufficiently large to assume equality in this approximation, then the performance of the discrete-time detector is the same as that of the continuous-time detector so long as the value of $d^2$ is fixed.

For n sufficiently large, then, one can mimic the log-likelihood ratio for the continuous-time case:

$$[dP^n_{\underline{S}+\underline{N}}/dP^n_{\underline{N}}](\underline{x}) \cong \ell^n(x)/\hat{A}^2_n(\underline{x}) - d^2/2\hat{A}^2_n(\underline{x}) \qquad (4\text{-}11)$$

where $\hat{A}^2_n(x) = n^{-1} \sum_{k=1}^{n} (\underline{x}*\underline{e}_k)^2/\lambda_k$,

$\ell^n(x) = \sum_{k=1}^{n} (\underline{x}*\underline{e}_k)(\underline{S}*\underline{e}_k)/\lambda_k$,

$\underline{R}\ \underline{e}_k = \lambda_k \underline{e}_k$, $k=1,\ldots,n$, and

$\{\underline{e}_k,\ k \geq 1\}$ is a complete orthonormal set in $E^n$.

The detector (4-11) has previously been given as an approximate likelihood ratio for large n by Picinbono and Vezzosi [13]. The above analysis indicates why this is so, and indicates how one can determine how large n must be in order to use the approximation.

The CFAP detector now becomes

$$\Lambda_{CFAP}(\underline{x}) = \ell^n(\underline{x})/\hat{A}_n(\underline{x}) \qquad (4\text{-}12)$$

where $\hat{A}_n(x) = [\hat{A}^2_n(\underline{x})]^{\frac{1}{2}}$.

For small n, one may wish to consider the CFAP detector given by

$$\bar{\Lambda}_{CFAP}(x) = \ell^n(\underline{x})/\sigma_n(\underline{x}) \qquad (4\text{-}13)$$

where $\sigma^2_n(\underline{x}) = (n-1)^{-1} \sum_{i=1}^{n} \left[\frac{(\underline{x}*\underline{e}_i)}{\lambda_i^{\frac{1}{2}}} - \frac{1}{n}\sum_{j=1}^{n} \frac{(\underline{x}*\underline{e}_j)}{\lambda_j^{\frac{1}{2}}}\right]^2$. Since the random variables $\{\underline{x}*\underline{e}_i/\sqrt{\lambda_i}:\ i=1,\ldots,n\}$ are i.i.d. $N(0,a^2)$ when A=a, and $\ell^n(\underline{x})$ is Gaussian, one may wish to assume that $\underline{S}^*_i\underline{e}_i/\sqrt{\lambda_i}$ = constant = $d/\sqrt{n}$ for $i=1,\ldots,n$. The test statistic divided by d then has a t distribution with n-1 degrees of freedom when noise only is present; this fact can be used to calculate $P_{FA}$. Under this same assumption, one can also obtain an expression for $P_D$ for this detector if the distribution of A is known, using the fact that $(n-1)\sigma_n{}^2$ is chi-square distributed with n - 1 degrees of freedom. One could use these considerations to determine a worst-case value of $P_D$ if the distribution of A is known to belong to a specified family, while maintaining a desired $P_{FA}$.

The problem of detecting a signal in Gaussian noise having unknown scale factor is familiar in active sonar. A detailed treatment of CFAP detection for this problem has been given by Grieve [7], who obtained CFAP optimality properties for (4-12).

## 5. APPLICATIONS TO DETECTION IN IMPULSIVE NOISE ENVIRONMENTS

As previously noted, Middleton's Class A univariate model is a special case of SIN. Detection in such noise has been analysed by several approaches. In [16], Spaulding and Middleton assume independent sampling and then develop bounds on likelihood ratio performance for communicating with a known signal over a channel in the presence of Class A noise. The Middleton model has also often been approximated by using only the first two or three terms: "Gauss-Gauss" or "Gauss-Gauss-Gauss" noise.

If the mixing random variable of the Middleton model remains constant over the observation interval, then the results given above can be used to provide detection results. It is not necessary to have independent sampling, but only to know the covariance matrix of the noise and the parameters U and $\Gamma$. For detection of a known signal, the preceding results can be used in several ways. They provide upper bounds on detection performance by giving the continuous-time detection performance. Secondly, they provide a method for obtaining exact detection performance for the discrete-time finite-sample-size detectors, and provide a means of calculating required sample size in order to simplify the detector structure. Thirdly, they can be used to obtain discrete-time CFAP detectors, as well as upper bounds on the performance of such detectors. Finally, the fact that the likelihood ratio detector can be implemented without knowing the distribution of the mixing r.v. A, once the sample size n is reasonably large (4-11), can provide a significant reduction of the complexity of the implementation. One need only adjust the threshold as a function of the parameters U and $\Gamma$, while the operation on the data is unchanged. Even this adjustment is not necessary if one is willing to use a CFAP detector.

The imbedding of the Middleton Class A model within the general SIN model thus provides a number of useful results. One may note the importance of the continuous-time model, which is often disregarded on the grounds that it is not relevant to practical signal detection. In the present case, the continuous-time model provides useful upper bounds on detection performance for both the likelihood ratio detector and the CFAP detectors. It also provides one with a practically-useful implementation and simplification of the apparently extremely-complicated discrete-time likelihood ratio detectors, and a rationale for making this simplification. The notion of orthogonal measures is central to these results.

## 6. EXTENSIONS TO THE SIN MODEL

The SIN model is not realistic for many situations, such as observation periods where the mixing r.v. A cannot be expected to take on a constant value. A more reasonable model in such situations would be generalized spherically-invariant noise, of the form $N(t) = A(t)G(t)$, where now $(A(t))$ is a stochastic process independent of the Gaussian process $(G(t))$. This reduces to a SIN model in the univariate case. Some work has previously been done for such a model [15], but general results are so far not available.


## ACKNOWLEDGEMENTS

This research was supported by ONR contracts N00014-81-K-0373, N00014-84-C-0212, and N00014-86-K-0039.

The authors thank H. Cherifi for helpful discussions and assistance.


## REFERENCES


1. C.R. Baker, Optimum quadratic detection of a random vector in Gaussian noise. <u>IEEE Trans. Commun.</u>, COM-14, 802-805 (1966).

2. C.R. Baker, On the deflection of a quadratic-linear test statistic, <u>IEEE Trans. Information Theory</u>, IT-15, 16-21 (1969).

3. C.R. Baker, On equivalence of probability measures, <u>Annals of Probability</u>, 1, 690-698 (1973).

4. C.R. Baker, Absolute continuity of measures on infinite-dimensional linear spaces, <u>Encyclopedia of Statistical Sciences</u>, 1, 3-11, Wiley, New York (1982).

5. C.R. Baker and A.F. Gualtierotti, Discrimination with respect to a Gaussian process, <u>Probability Theory and Related Fields</u>, 71, 159-182 (1986).

6. R. Dwyer, A technique for improving detection and estimation of signals contaminated by underice noise, <u>J. Acoustical Soc. America</u>, 74, 124-130 (1983).

7. P.G. Grieve, The optimum constant false alarm probability detector for relatively coherent multichannel signals in Gaussian noise of unknown power, <u>IEEE Trans. Information Theory</u>, IT-23,

8. S. Geman, An application of the method of sieves: functional estimator for the drift of a diffusion, Reports in Pattern Analysis, 92, Div. Applied Math., Brown Univ. (1980).

9. J. Keilson and F.W. Steutel, Mixtures of distributions, moment inequalities, and measures of exponentiality and normality, Annals of Probability, 2, 112-130 (1974).

10. T. Hida, Canonical representations of Gaussian processes and their applications, Mem. Coll. Science, Univ. Kyoto, 33A, 109-155 (1960).

11. R.S. Liptser and A.N. Shiryayev, Statistics of Random Processes I. General Theory, Springer-Verlag, New York (1977).

12. D. Middleton, Statistical-physical models of electromagnetic interference, IEEE Trans. on Electromagn. Compat., EMC-19, 106-127 (1977).

13. B. Picinbono and G. Vezzosi, Détection d'un signal certain dans un bruit non stationnaire et non Gaussien, Annales des Telecommunications, 25, 433-439 (1970).

14. C.R. Rao and V.S. Varadarajan, Discrimination of Gaussian processes, Sankhyā, 25A, 303-330 (1963).

15. A.D. Spaulding, Locally optimum and suboptimum detector performance in a non-Gaussian interference environment, IEEE Trans. Commun., COM-33, 509-517 (1985).

16. A.D. Spaulding and D. Middleton, Optimum reception in an impulsive interference environment - Part I: Coherent detection; Part II: Incoherent detection, IEEE Trans. Commun., COM-25, 910-934 (1977).

17. R.L. Spooner, On the detection of a known signal in a non-Gaussian noise process, J. Acoustical Soc. America, 44, 141-147 (1968).

18. K. Yao, A representation theorem and its application to spherically-invariant random processes, IEEE Trans. Information Theory, IT-19, 600-608 (1973).

CHAPTER 7

# DETECTION AND CONTRAST

B. Picinbono and P. Duvaut

## 1. INTRODUCTION

It is well known that the **likelihood ratio** is a sufficient statistic for the test between two simple hypotheses, which is the most basic and elementary detection problem [14] [25]. Moreover any monotonic function of this likelihood ratio gives an equivalent receiver in the sense that the performance is the same. This property is very often used by the introduction of the log-likelihood ratio. In other words there is no one optimal receiver but a class of equivalence of optimal receivers.

But in many practical detection problems the likelihood ratio test cannot be implemented. This is sometimes due to its complexity and sometimes to lack of knowledge of the probability distributions which makes its exact calculation impossible.

In this case it is possible to introduce other measures of quality for signal detection, and the **contrast** appears as a good candidate. It is a particular case of second-order measure of classification studied by Gardner [12] and has also a certain relationship with the concept of general signal to noise ratio (see also [12]). Indeed, in a very wide sense, the aim of any detector is to realize the maximum of contrast between the two situations (noise only and signal plus noise) in order to obtain the best detection performance.

Apparently there is no relation between the contrast and the statistical approaches of detection problems, but we will see that this is not exactly the case and one of the purposes of this paper is to discuss this point. Moreover we will see that the contrast is non invariant in a monotonic transformation of the receiver output, and then the problem of finding the receiver equivalent to the likelihood ratio and giving the maximum contrast appears.

After presenting the possible definitions of the contrast we develop some of its basic properties. In particular we show how singular detection is connected with infinite contrast. By introducing an appropriate scalar product we also show that the research of the maximum contrast is equivalent to a minimum distance problem which can be solved by geometrical techniques. Some examples of such optimization problems are given. Finally we discuss in the last section the interest of the contrast criterion in quantized detection. Indeed if the threshold of the receiver is quantized, the monotonic invariance of the receiver performance is no longer true and we show in some examples how the contrast criterion can throw light on this problem.

## 2. DEFINITIONS OF THE CONTRAST

Let us call $\mathbf{x}$ the observation vector which is a real random vector of $\mathbf{R}^N$. Its probability distributions under the two hypotheses are $p_0(\mathbf{x})$ and $p_1(\mathbf{x})$. In order to make a decision from the observation vector $\mathbf{x}$ we must process this vector by a system which calculates the function $S(\mathbf{x})$. Many different terms are used for this function : classifier [12], statistic [14], p.95 and in the following we will use the term of filter, well adapted to engineering literature, after noticing that this filter has no reason, in general, to be linear.

The optimal filter in the statistical sense is of course the likelihood ratio, (LR), $L(\mathbf{x}) = p_1(\mathbf{x})/p_0(\mathbf{x})$ which is a sufficient statistic for the problem of two hypotheses testing (see [14], p.95).

The **contrast $C_\pi$ of a filter $S(x)$** is defined by

$$C_\pi(S) = C[S;\pi] \triangleq [E_1(S) - E_0(S)]^2/V_\pi(S), \qquad (2\text{-}1)$$

where $E_0$ and $E_1$ are respectively the expectation values under $H_0$ and $H_1$ and $V_\pi$ the variance of $S(\mathbf{x})$ corresponding to the probability distribution defined by

$$p_\pi(\mathbf{x}) \triangleq (1-\pi)\, p_0(\mathbf{x}) + \pi\, p_1(\mathbf{x}). \qquad (2\text{-}2)$$

This distribution describes a mixture of the two densities characterized by the mixing probability $\pi$, $(0 \leq \pi \leq 1)$.

The contrast has obvious relations to some other detection

criteria already introduced.

If $\pi = 0$, the contrast is the classical **deflection criterion**, [15], p.161-163, [2].

If $\pi = 1/2$, it corresponds to a definition of a signal to noise ratio introduced by Rudnick [22].

Finally, if $\pi$ is the a priori probability of the hypothesis $H_1$, $C_\pi(S)$ is related to the general signal to noise ratio introduced by Gardner in [12] and defined by

$$C'_\pi(S) \triangleq [E_1(S) - E_0(S)]^2 / [\pi V_1(S) + (1-\pi) V_0(S)], \qquad (2\text{-}3)$$

where $V_0$ and $V_1$ are respectively the variances under $H_0$ and $H_1$. Indeed, after a simple calculation, we obtain

$$C'_\pi = C_\pi / [1 - \pi(1-\pi) C_\pi] \qquad (2\text{-}4)$$

In the following discussion we will see that it is simpler to work with $C_\pi$ than with $C'_\pi$.

Finally we notice that the contrast is directly related to the **asymptotic relative efficiency (ARE)** very often used in detection literature [14], p.228, [5], [20]. For this purpose let us suppose that $H_1$ corresponds to the presence of a deterministic signal $\alpha\, \mathbf{s}$. Then $p_1(\mathbf{x})$ becomes $p_0(\mathbf{x}-\alpha\mathbf{s})$. If $\alpha$ is small, we can neglect the terms in $\alpha^k$, k>1, which allows us to write

$$E_1(S) = E_0(S) + \alpha E_0(\mathbf{s}^T \nabla S). \qquad (2\text{-}5)$$

Then the contrast $C_0(S)$ becomes

$$C_0(S) = \alpha^2 \eta \qquad (2\text{-}6)$$

$$\eta \triangleq [E_0(\mathbf{s}^T \nabla S)]^2 / V_0(S). \qquad (2\text{-}7)$$

In the case of a constant signal s in a white noise, the optimum filter $S(\mathbf{x})$ (monotonic function of the L.R.) can be expressed in the form $\sum g(x_i)$, which gives

$$\eta = [E_0(g')]^2 / V_0(g), \qquad (2\text{-}8)$$

which is the classical definition of the ARE.

Finally it is worth noticing some particular expressions of the contrast.

At first let us consider the case of a linear filter $S(\mathbf{x}) = \mathbf{h}^T\mathbf{x}$. Taking

$$\mathbf{s} \triangleq E_1(\mathbf{x}) - E_0(\mathbf{x}) \tag{2-9}$$

$$K_\pi \triangleq E_\pi\{[\mathbf{x} - E_\pi(\mathbf{x})]\,[\mathbf{x} - E_\pi(\mathbf{x})]^T\}, \tag{2-10}$$

we get

$$C_\pi(\mathbf{h}) = (\mathbf{h}^T\,\mathbf{s})^2/\mathbf{h}^T\,K_\pi\,\mathbf{h}. \tag{2-11}$$

This contrast is of course zero if $E_0(\mathbf{x}) = E_1(\mathbf{x})$. On the other hand it is maximum for the matched filter defined by

$$\mathbf{h}_\pi = K_\pi^{-1}\,\mathbf{s} \tag{2-12}$$

and the maximum value of $C_\pi$ is

$$d_\pi^2 = \mathbf{s}^T\,K_\pi^{-1}\,\mathbf{s}\ . \tag{2-13}$$

Secondly let us consider the case of a test characteristic function (t.c.f.) $\phi(\mathbf{x})$. It is a function which has only two possible values 0 or 1 and then satisfies the relation $\phi^2 = \phi$. As a consequence its variance is

$$V_\pi(\phi) = E_\pi(\phi) - E_\pi^2(\phi)\ , \tag{2-14}$$

with

$$E_\pi(\phi) = (1 - \pi)\,E_0(\phi) + \pi\,E_1(\phi). \tag{2-15}$$

But $E_0(\phi)$ and $E_1(\phi)$ can respectively be interpreted as the false alarm and detection probabilities $\alpha$ and $\beta$, which gives

$$C_\pi(\phi) = [\beta(\phi) - \alpha(\phi)]^2\{(1-\pi)\,\alpha(\phi) + \pi\beta(\phi) - [(1-\pi)\,\alpha(\phi) + \pi\beta(\phi)]^2\}^{-1}\ . \tag{2-16}$$

The most interesting case appears for the deflection criterion ($\pi = 0$) which gives

$$C_0(\phi) = [\beta(\phi) - \alpha(\phi)]^2\,[\alpha(\phi) - \alpha^2(\phi)]^{-1}\ . \tag{2-17}$$

Finally if the receiver is a threshold receiver, associated to a filter $S(\mathbf{x})$, which means that $\phi(\mathbf{x}) = 1$ if and only if $S(\mathbf{x}) > t$, the t.c.f. is defined by $t$, or also by the corresponding false alarm $\alpha$, and the previous equation becomes

$$C_0(\alpha) = [\beta(\alpha) - \alpha]^2 \, [\alpha - \alpha^2]^{-1} , \tag{2-18}$$

where $\beta(\alpha)$ is the **Receiver Operating Characteristic** (or ROC) curve (see [14] p.88 and 107).

## 3. PROPERTIES OF THE CONTRAST

In this section are presented some important properties of the contrast which are used in the following discussion.

### 3.1. Existence of the contrast. Singular detection

From (2-1) it is clear that the contrast of a filter S is defined if S is a second order random variable under $H_0$ and $H_1$ and if its variance $V_\pi$ is not equal to zero. This variance can be written

$$V_\pi = \pi V_1 + (1-\pi) V_0 + \pi(1-\pi) (m_1 - m_0)^2 , \tag{3-1}$$

where $V_0$, $m_0$, $V_1$, $m_1$ are respectively the variances and mean values of S under $H_0$ and $H_1$.

For $\pi \neq 0$ and $\pi \neq 1$ this variance can be equal to zero only if $V_0 = V_1 = 0$ and $m_0 = m_1$. In this case the contrast is not defined, as a ratio of two terms equal to zero. But as $S(\mathbf{x})$ has almost surely the same value under $H_0$ and $H_1$ then no detection is possible and we can assume that the contrast is vanishing. For $\pi = 0$ (deflection criterion) the variance $V_\pi$ is equal to zero if $V_0 = 0$ which means that $S(\mathbf{x})$ is almost surely equal to $m_0$. If $m_1 = m_0$ the contrast is not defined while if $m_1 \neq m_0$ it is infinite.

It is interesting to discuss the problem of singular detection in terms of contrast. The detection problem is singular if it is possible to find a test function $\phi(\mathbf{x})$ such that the false alarm and detection probabilities are respectively equal to 0 and 1. This can be written

$$\alpha = \int \phi\,(\mathbf{x})\; p_0(\mathbf{x})\; d\mathbf{x} = 0 \tag{3-2}$$

$$\beta = \int \phi\,(\mathbf{x})\; p_1(\mathbf{x})\; d\mathbf{x} = 1 \tag{3-3}$$

We deduce from (2-16) that $C_\pi = [\pi(1-\pi)]^{-1}$ which is infinite for $\pi = 0$ and $\pi = 1$. On the other hand it results from (2-4) that $C'_\pi$ defined by (2-3) is infinite for any value of $\pi$.

Conversely if there exists a filter $S(\mathbf{x})$ for which $m_0 \neq m_1$, and which gives an infinite contrast for $\pi = 0$ and $\pi = 1$, then the detection problem is singular. Indeed, as the numerator of (2-3) is $(m_1 - m_0)^2$ which is not equal to zero, we deduce that $V_0 = V_1 = 0$, which means that $S(\mathbf{x})$ is almost surely equal to $m_0$ or $m_1$ respectively under $H_0$ and $H_1$. If $D_0$ and $D_1$ are the subsets of $\mathbf{R}^N$ where $S(\mathbf{x})$ is equal respectively to $m_0$ or $m_1$, we have $D_0 \cap D_1 = \emptyset$ and $p_0(\mathbf{x}) = 0$ if $\mathbf{x} \in D_1$, $p_1(\mathbf{x}) = 0$ if $\mathbf{x} \in D_0$. That gives $p_0(\mathbf{x}).p_1(\mathbf{x}) = 0$, which is the definition of singular detection.

### 3.2. Invariance of the contrast

From (2-1) we deduce immediately that

$$C_\pi (\lambda S + \mu) = C_\pi(S), \lambda \neq 0. \tag{3-4}$$

Then to any filter S we can associate a class $C_S$ of filters equivalent to S and obtained by the operation $\lambda S + \mu$, $\lambda \neq 0$.

Similarly to any family of filters we can associate its **extension** $\tilde{F}$ by the operation $\lambda S + \mu$, $S \in F$, and if $S \in \tilde{F}$, then $\lambda S + \mu \in \tilde{F}$, $\lambda \neq 0$.

Moreover if we introduce

$$m_\pi \triangleq E_\pi(S), \tag{3-5}$$

which is the expectation value of S under the distribution (2-2), all the filters $\lambda \overline{S}$ where

$$\overline{S} \triangleq S - m_\pi \tag{3-6}$$

are elements of a subclass $\overline{C}_S$ of $C_S$ which is the subclass of zero mean filters equivalent to S.

Finally to any filter we can associate one filter, element of $\overline{C}_S$ and such that $V_\pi(S) = 1$. This filter $\overline{S}_n$ is **the zero mean and normalized filter equivalent to S** and it will play an important role in the following discussion.

### 3.3. Contrast and scalar product

The scalar product of two filters $u(\mathbf{x})$ and $v(\mathbf{x})$ is defined by

$$\langle u,v\rangle_\pi \triangleq E_\pi[u(\mathbf{x})\ v(\mathbf{x})] = \int u(\mathbf{x})\ v(\mathbf{x})\ p_\pi(\mathbf{x})\ d\mathbf{x}\ , \tag{3-7}$$

where $p_\pi(\mathbf{x})$ is given by (2-2). This product is well defined if $u(\mathbf{x})$ and $v(\mathbf{x})$ are second order random variables under the distribution $p_\pi(\mathbf{x})$.

With this definition we will express the contrast as a ratio of scalar products, which is very interesting for the following discussion.

At first the denominator of (2-1) can be written

$$V_\pi(S) = \langle\ \overline{S},\ \overline{S}\ \rangle_\pi\ , \tag{3-8}$$

where $\overline{S}$ is defined by (3-6).

Let us now consider the numerator N of (2-1). It is given by

$$N^{1/2} = \int \overline{S}(\mathbf{x})[p_1(\mathbf{x}) - p_0(\mathbf{x})]\ d\mathbf{x}\ , \tag{3-9}$$

and using the filter defined by

$$\begin{aligned} R_\pi(\mathbf{x}) &\triangleq [p_1(\mathbf{x}) - p_0(\mathbf{x})].\ [\pi p_1(\mathbf{x}) + (1-\pi)\ p_0(\mathbf{x})]^{-1} \\ &= [L(\mathbf{x}) - 1]\ .\ [\pi\ L(\mathbf{x}) + 1-\pi]^{-1}\ , \end{aligned} \tag{3-10}$$

where $L(\mathbf{x})$ is the likelihood ratio (LR), we get

$$N^{1/2} = \langle\ \overline{S},\ R_\pi\ \rangle_\pi\ . \tag{3-11}$$

Finally the contrast of S is

$$C_\pi(S) = C_\pi(\overline{S}) = [\langle\overline{S},R_\pi\rangle_\pi]^2\ /\ \langle\overline{S},\overline{S}\rangle_\pi\ . \tag{3-12}$$

Let us illustrate this expression by some simple examples. If $\pi=0$

$$R_0(\mathbf{x}) = L(\mathbf{x}) - 1 \tag{3-13}$$

and as

$$\langle\overline{S},\ L-1\rangle_0 = \langle\overline{S},\ L\rangle_0 \tag{3-14}$$

we get

$$C_0(S) = [\langle\overline{S},\ L\rangle_0]^2\ /\ \langle\overline{S},\overline{S}\rangle_0\ . \tag{3-15}$$

That is the expression of the deflection criterion associated to a filter $S(\mathbf{x})$ in terms of the LR $L(\mathbf{x})$. As indicated in the introduction this expression establishes the relation between the statistical decision theory and the contrast (or deflection) approach of detection problems. This relation is discussed in more detail in the following.

If $\pi = 1/2$, which is the case considered by Rudnick [22], we obtain

$$R(\mathbf{x}) = 2[L(\mathbf{x}) - 1] \cdot [L(\mathbf{x}) + 1]^{-1} . \tag{3-16}$$

### 3.4. Contrast and Hilbert space of filters

The Hilbert space of filters is the space H of functions $u(\mathbf{x})$ in which the scalar product is defined by (3-7). The elements of this space must have a finite norm, which means that $u(\mathbf{x})$ is a second order random variable under the distribution $p_\pi(\mathbf{x})$. In particular the condition which secures that $R_\pi(\mathbf{x})$ defined by (3-10) is in this space is

$$\int \{p_1(\mathbf{x}) - p_0(\mathbf{x})\}^2 \{p_0(\mathbf{x}) + \pi[p_1(\mathbf{x}) - p_0(\mathbf{x})]\}^{-1} \, d\mathbf{x} < \infty . \tag{3-17}$$

We assume in the following that this condition is satisfied. Moreover we can notice that for the deflection criterion ($\pi = 0$) it is equivalent to assume that the LR has a finite mean value under $H_1$.

**Hilbert subspace of zero mean filters**

Consider the subspace $\overline{H}$ consisting of H of filters $S(\mathbf{x})$ such that $E_\pi(\overline{S}) = 0$, which can be written in the form

$$\langle \overline{S}, 1 \rangle_\pi = 0 . \tag{3-18}$$

This subspace is orthogonal to the filter giving the output 1 to any input $\mathbf{x}$. Moreover the operation described by (3-6) is a projection on H, and

$$\overline{S} \triangleq S - E_\pi(S) = \mathrm{Proj}[S|\overline{H}] . \tag{3-19}$$

In all the following we will work only in $\overline{H}$. It is also important to notice that $R_\pi(\mathbf{x})$ is an element of $\overline{H}$, which is an important property for the following discussion.

**Family of normalized filters**

Consider the family of filters $S_n$ such that $\langle S_n, S_n \rangle_\pi = 1$. All the elements of this family belong to the unit sphere of H. Moreover to any filter S(**x**) we can associate one filter, element of this family, and defined by

$$S_n(\mathbf{x}) = \langle S,S \rangle_\pi^{-1/2} \; S(\mathbf{x}) \; . \qquad (3\text{-}20)$$

Using these definitions, the contrast $C_\pi(S)$ given by (2-1) and (3-12) can be written in the form

$$C_\pi(S) \quad = \quad R_\pi^{-2} \, C_n(S) \qquad (3\text{-}21)$$

$$R_\pi^2 \quad = \quad \langle R_\pi, R_\pi \rangle_\pi \qquad (3\text{-}22)$$

$$C_n(S) \quad \triangleq \quad \langle \overline{S}_n, R_n \rangle_\pi^2 \quad . \qquad (3\text{-}23)$$

Then the contrast is proportional to the normalized contrast $C_n(S)$ defined by (3-23). In this equation $\overline{S}_n(\mathbf{x})$ is the filter deduced from S(**x**) by using (3-6) and (3-20), which means the zero mean normalized filter equivalent to S. Similarly $R_n$ is deduced from $R_\pi$ defined in (3-10) by application of (3-20).

These equations are the most compact expressions of the contrast and moreover have a **simple geometrical interpretation**. Indeed as $\overline{S}_n$ and $R_n$ are vectors of the unit sphere of $\overline{H}$, their scalar product is given by

$$\langle \overline{S}_n, R_n \rangle = \cos \alpha(\overline{S}_n, R_n) \; , \qquad (3\text{-}24)$$

where $\alpha(\overline{S}_n, R_n)$ is the angle between these two vectors. This geometrical interpretation is particularly convenient for solving optimization problems.

## 4. OPTIMIZATION PROBLEMS

Using the contrast criterion the optimum receiver is the filter giving its maximum value. This optimum filter is of course strongly depending on the class of filters used in each particular detection problem, and we begin by establishing a general result before discussing some particular cases of applications.

## 4.1. General result

Let us consider an arbitrary family F of regular filters. This expression means that all the elements of this family have a finite norm or that F is a subset of H defined in 3.4.

The problem discussed in this section is to find the elements of F giving the maximum contrast and called **optimum filters** in the family F.

For this purpose we associate to F its extension $\tilde{F}$ defined in 3.2. Moreover $\tilde{F}_{0,n}$ is the subset of $\tilde{F}$ containing only zero mean normalized filters. In other words

$$\tilde{F}_{0,n} = \tilde{F} \cap \bar{H} \cap US , \qquad (4\text{-}1)$$

where $\bar{H}$ is defined in 3.4 and US is the unit sphere of H.

It is worth noticing that to each filter S of F there corresponds two equivalent filters of $\tilde{F}_{0,n}$ which are $\bar{S}_n$ and $-\bar{S}_n$.

**Proposition.** Let us suppose that R is regular, which is secured by (3-17), and let us consider a family F of regular filters. Then the filters of F giving the maximum contrast are equivalent to the elements of $\tilde{F}_{o,n}$ such that their distance to $R_n$ is minimum.

**Proof.** As $\bar{S}_n$ and $-\bar{S}_n$ belong to $\tilde{F}_{o,n}$, we can consider only the elements of $\tilde{F}_{o,n}$ for which $\langle \bar{S}_n, R_n \rangle_\pi \geq 0$. Moreover the square of the distance between $\bar{S}_n$ and $R_n$ is

$$d^2 = 2 \; [1 - \langle \bar{S}_n, R_n \rangle_\pi] \qquad (4\text{-}2)$$

and the maximum of $C_n(S)$ given by (3-23) when $\langle \bar{S}_n, R_n \rangle_\pi \geq 0$ is obtained when $d^2$ is minimum.

## 4.2. Interpretation and direct consequences

### a. Importance of the use of $\tilde{F}_{o,n}$

It is worth noticing the importance of the family in which the minimum distance is searched. In general the optimum filter giving the maximum contrast is not the element of F giving the minimum distance to $R_n$. That is also in general not even the case for $F_{o,n}$. Indeed, as indicaded just above, to each element of $F_{o,n}$ correspond two equivalent filters of $\tilde{F}_{o,n}$ and it is perfectly possible that the minimum distance for $\tilde{F}_{o,n}$ is the maximum distance in $F_{o,n}$.

**b. Relations between contrast and statistical decision theories**

Let us suppose that the filter $R(\mathbf{x})$ belongs to the family F, which is in particular the case when there is no constraint and means that F is the Hilbert space H defined in 3.4. In this case it is obvious that the optimum filter of the family F in terms of contrast is $R(\mathbf{x})$ itself.

In particular if we use the deflection criterion ($\pi$=0), $R(\mathbf{x})$ is given by (3-13) and as $L(\mathbf{x})$ and $L(\mathbf{x})-1$ are equivalent in terms of contrast, the optimum filter is $L(\mathbf{x})$. **Then the filter which maximises the deflection criterion is the likelihood ratio receiver.**

This result is very interesting because it establishes a relation between two apparently completely different approaches of decision theory. It was indicated under stronger conditions by Poor [21], and also partially by Rudnick [22]. Another proof can also be found in [19].

**c. Differences between contrast and statistical decision theories**

Nevertheless there is a strong difference between the two approaches. In the statistical decision theory the LR $L(\mathbf{x})$ is compared to a threshold to make the decision. In other words any receiver using a monotonic function of $L(\mathbf{x})$ is equivalent to $L(\mathbf{x})$ and this property is used typically by taking the logarithm of $L(\mathbf{x})$.

This "monotonic property" is no longer true with the contrast, because it is only invariant by the transformations studied in 3.2. This point was indicated in [19] and will also be discussed in the following.

**d. Equivalent receivers**

Let us consider a filter $S(\mathbf{x})$. All the filters on the form $\lambda\, S(\mathbf{x}) + \mu$ give the same contrast as $S(\mathbf{x})$ and form the class of equivalence of $S(\mathbf{x})$ in terms of contrast. On the other hand all the filters deduced from $S(\mathbf{x})$ by a monotonic transformation and used with a threshold decision rule give the same performance in terms of Receiving Operating Characteristics and are equivalent in terms of ROC properties. It is obvious that the contrast class of equivalence is a subclass of the ROC class of equivalence.

**e. Optimal receiver and Hilbert space constraints**

If the family F is an Hilbert subspace of H, then its extension $\tilde{F}$ is also such an Hilbert subspace and the minimum distance element is obtained by projection and can be written on the form

$$S(\mathbf{x}) = \text{Proj}\ [R_\pi(\mathbf{x})|\tilde{F} \cap \overline{H}] = \text{Proj}\ [R_\pi(\mathbf{x})|\tilde{F}], \qquad (4\text{-}3)$$

because, as noticed previously, $R_\pi$ is an element of $\overline{H}$.

It is worth noticing the importance of $\tilde{F}$, and it is easy to check that the projection of $R_\pi(\mathbf{x})$ directly on the Hilbert space F can be completely different.

As the projection appearing in (4-3) is a linear operation, this result shows that each problem of optimal filtering in a family of filters which has an Hilbert subspace structure **can be solved by a linear method.**

For example the family of **linear filters** $\mathbf{h}^T\ \mathbf{x}$ is an Hilbert subspace and the matched filter (2-12) can be found as a projection of $R_\pi(\mathbf{x})$. Particularly if the deflection criterion is used, this shows that the classical matched filter $K_0^{-1}\ \mathbf{s}$ is the projection of the likelihood ratio. The same result appears if a linear constraint, as for example a constraint of Noise Alone Reference is imposed in the filter [18], [10], [9].

The same situation appears in the case of a family of **quadratic filters**, and in the Gaussian case the optimum filtering obtained by projection is the Eckart filter introduced in [8] and discussed in [2] and in [11], [26], [23], [4], [16], [1]. But this filter is obtained only with the assumption of a Gaussian noise, while the projection method can be applied for any probability distribution. Examples of such results will be published in another paper.

f. **Test characteristic functions**

Let us suppose that the family F in consideration is the class of test characteristic functions, i.e., of filters giving only two possible values 0 or 1, introduced at the end of the section 2. Of course this family is not a Hilbert space and the optimal element cannot be obtained by projection. In order to simplify the discussion we assume, as often previously, that the deflection criterion is used, which means that $\pi = 0$. In other words the problem is to find the t.c.f. giving the best approximation of the likelihood ratio in the sense of minimum distance.

For this purpose let us call $T_\alpha$ the class of all t.c.f. giving a false alarm probability equal to $\alpha$ and a detection probability greater than $\alpha$. We deduce from (2-18) that in this class the t.c.f. giving the contrast maximum is also the t.c.f. giving the maximum of detection probability, and that is the Neymann-Pearson test comparing the LR to a threshold determined by $\alpha$. Then for a given value of $\alpha$, the maximum of the contrast is given by (2-18), where $\beta(\alpha)$ is calculated with the

LR test. Now it remains to search the value of $\alpha$ giving the maximum of $C_o(\alpha)$ and we will discuss this point in an example.

Let us consider the detection problem of a deterministic signal **s** in a Gaussian noise with the distribution N(**0**, K). The L.R. is equivalent to the matched filter which calculates $y = \mathbf{h}^T \mathbf{x}$ and the distributions of y under $H_0$ and $H_1$ are respectively $N(0,d^2)$ and $N(d^2,d^2)$. These distrubtions become N(0,1) and N(d,1) if we use $d^{-1}$ y instead of y, which is completely equivalent.

The maximum contrast given by (2-18) is expressed as a function of $\alpha$. It can also be expressed as a function of the threshold which gives

$$C_o(t) = [\beta(t)-\alpha(t)]^2 \, [\alpha(t) - \alpha^2(t)]^{-1} \, , \tag{4-4}$$

where

$$\alpha(t) = \int_t^\infty n(\theta)\, d\theta; \quad \beta(t) = \int_t^\infty n(\theta-d)\, d\theta, \tag{4-5}$$

and n(u) is the probability density associated to N(0,1). Then this contrast is a function of t and d which is represented in figure 1.

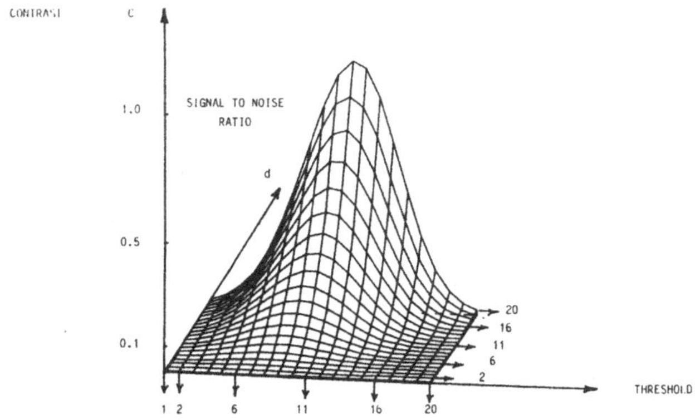


Figure 1 : Representation of the contrast $C_o(t,d)$.

This figure shows that for each value of the signal to noise ratio d, there exists an optimum threshold for which the contrast is maximum. Moreover we see that the contrast tends to zero for extreme values of the threshold and this property can be deduced from the

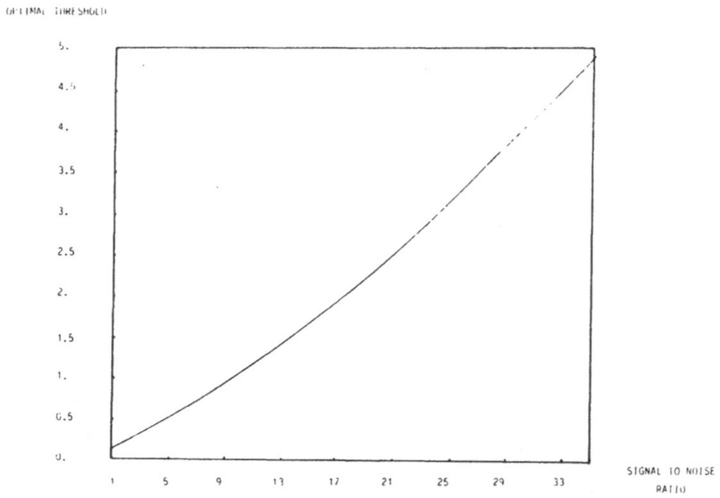


Figure 2 : Optimum threshold corresponding to the contrast maximum function of the signal to noise ratio.

asymptotic behaviour of the error function. Using the Neyman-Pearson point of view we can also say that for each value of d there exists a false alarm probability $\alpha(d)$ corresponding to the maximum of the contrast, and this is given in figure 2.

We can see in figure 2 that the optimum threshold is an increasing function of d, Signal to Noise Ratio. Actually, this result means that, as d increases, the highest Contrast is reached with lower risk, taking into account that the optimal false alarm probability is decreasing. From a physical point of view this conclusion was expected, noting that as d increases, the right decision must be taken with less risk.

**g. Additional comments**

If the family F has no particular geometrical structure there is no general method for solving the optimization problem, which remains a minimum distance problem.

Finally we can indicate that a similar result was indicated by

Czarnecki ([7], [6]) under stronger conditions. Instead of using the general definition of the contrast, only the locally optimum receiver for a constant signal in white noise was used. Then the measure of performances was the efficacy $\eta$ given by (2-8) and we have seen in section 2 that it is a particular case of the contrast. Moreover the family $\tilde{F}$ was not systematically introduced, and replaced by "a simple normalisation of the elements of F", which is not strictly equivalent.

## 5. CONTRAST AND QUANTIZED DETECTION

As indicated in the introduction and discussed in 4.2. c and d any threshold receiver has the monotonic invariance property, which means that the performance is unchanged if the output of the receiver is transformed by any monotonic function, provided that the threshold is also transformed by the same function.

The true reason for this monotonic invariance property is the fact that it is assumed that the decision threshold is a **pure mathematical threshold** using the inequalities $S(\mathbf{x}) > t$ or $S(\mathbf{x}) \leq t$. But in most practical systems, numerical devices are used, which means that the threshold is quantized. In this case the monotonic invariance has no reason to remain true, and it is important to discuss this point in more detail now.

As we will see, threshold quantization can introduce dramatic degradation of the performance, and we will show that this point can also be discussed in terms of contrast. Apparently this problem was not seriously considered in the past and the only connected work is presented in [17].

### 5.1. **Detection with a quantized threshold**

Let us call X the random variable, output of a decision system, before quantization, and F(x) its distribution function [$F(x)=Pr(X<x)$]. In order to calculate the ROC curves we need these distributions under $H_0$ and $H_1$. **Before quantization** the alarm probability related to a threshold t is

$$\pi_l(t) : 1 - F_l(t), \; l = 0,1 \tag{5-1}$$

In order to calculate the same probabilities **after quantization,** it is necessary to use a model for this operation. Many models are possible [3], p.47, but in this paper we will restrict our attention to a special case of **random quantization.** Of course the following calculations can also be performed with other models of quantization, and even if the details are very different, the main conclusions are the same.

Let us call Y the output of the quantizer. We assume that

$$Y = X + Z , \tag{5-2}$$

where the random variable Z describes the quantization noise. In order to simplify the calculation we assume also that X and Z are independant and that the probability density of Z is

$$p(z) = \frac{1}{2} [\delta(z+q/2) + \delta(z-q/2)], \tag{5-3}$$

where $\delta(\ )$ is the Dirac distribution and q the quantization step. As a consequence the alarm probability after quantization becomes

$$\pi_{q,l}(t) = \frac{1}{2} [\pi_l(t-q/2) + \pi_l(t+q/2)], \quad l = 0, 1, \tag{5-4}$$

where $\pi_l$ is given by (5-1). This quantization procedure can be interpreted on the following way : if $X < t-q/2$ or $X \geq t + q/2$, we decide respectively $H_0$ or $H_1$. On the other hand if $t-q/2 \leq X < t+q/2$, we decide $H_0$ or $H_1$ with the same probability 1/2. Of course if $q = 0$, the effect of quantization disappears.

The detection problem with quantization remains the same as without quantization. In other words it consists in finding the optimal filter giving the best performance, or for a given false alarm probability the greater detection probability. But of course, as these probabilities are calculated after quantization, it is no longer possible to apply the Neyman Pearson criterion and then there is no reason to find the L.R. as optimal filter. Moreover it is evident that the monotonic invariance property disappears also.

To check all these points, and to compare with the contrast approach we will consider as in 4.2.f. the most basic detection problem of a deterministic signal **s** in a zero mean Gaussian noise with known covariance matrix.

As indicated in 4.2.f. the decision is taken by considering a random variable X which has under $H_0$ and $H_1$ respectively the distributions N(0,1) and N(d,1), where $d^2$ is the signal to noise ratio.

The purpose of the following calculation is to discuss the effect of a monotonic transformation on X first in terms of contrast, secondly in terms of quantized detection, and to compare the results.

## 5.2. Behaviour of the contrast in a particular monotonic transformation.

Our purpose is to introduce a particular class of monotonic transformations of X which depend only on a limited number of parameters in order to simplify the discussion.

This transformation is defined by

$$S_a(x) = k\ Sg(x)\ [1-e^{-a\ Sg(x)}], \qquad (5-5)$$

where $Sg(x) = x/|x|$ . It is clear that it is a monotonic transformation depending only on two parameters a and k.

As a matter of fact we will suppress this last constant by assuming that under $H_0$ the variance of $S_a(x)$ is one. Indeed as the signal following $S_a(x)$ will be quantized with a step q, it appears that the physical parameter is the ratio between $q^2$ and the variance of $S_a(x)$ under $H_0$, and the simplest way is to assume that $S_a(x)$ has an unit variance. This determines k by

$$k = [1-4\exp(\frac{a^2}{2})\ \mathrm{erfc}(a)+2\exp(2a^2)\ \mathrm{erfc}(2a)]^{-1/2} \qquad (5-6)$$

where erfc is defined by

$$\mathrm{erfc} = (2\pi)^{-1/2} \int_x^\infty \exp(-\frac{1}{2}u^2)\ du. \qquad (5-7)$$

Finally it is worth noticing that if in (5-5), $a \to 0$, then $S_a(x)$ becomes the identical transformation, and if $a \to \infty$ it becomes proportional to the sign receiver which is often used in the non-parametric detection context [24] [13].

The function $S_a(x)$ exhibits a saturation effect, in such a way that the contrast decreases when a increases. As a is taking a wide set of values, many situations may occur at the output with variable contrast as well, keeping the monotonic property.

But of course the contrast can change strongly and we begin at first by its calculation. Starting from (2-3) and assuming that $\pi = 0$, (deflection criterion) and noticing that with (5-6), $V_o = 1$ we find

$$C[S_a]=k^2e^{-d^2}.\left[2e^{d^{2/2}}.\mathrm{erf}(d)-e^{-\frac{(a-d)^2}{2}}\mathrm{erfc}(a-d)+e^{-\frac{(a+d)^2}{2}}\mathrm{erfc}(a+d)\right]^2, \qquad (5-8)$$

where erf(x) = 1/2 - erfc(x). This contrast is depending on the parameter a and the signal to noise ratio at the output of the matched-filter d. The function (5-8) is drawn in 3 dimensions in figure 3.

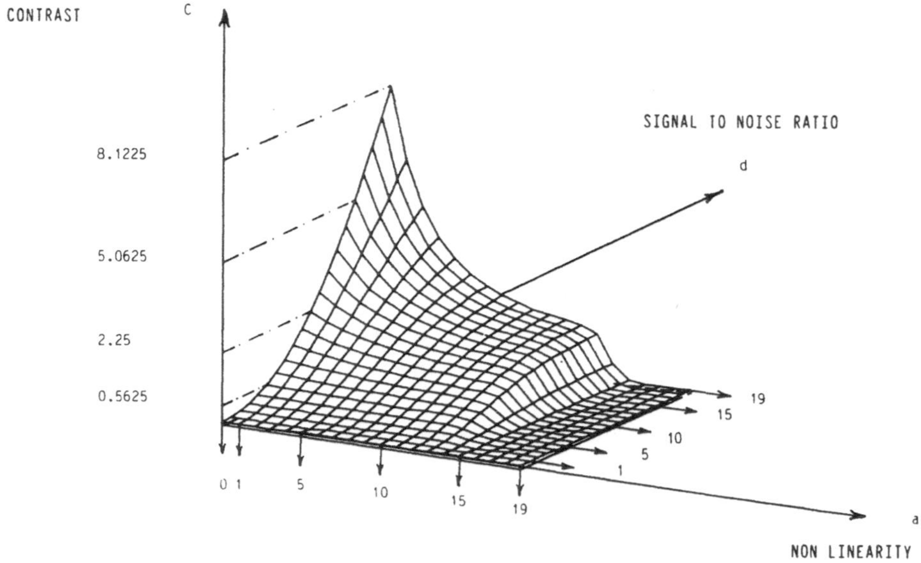


Figure 3 : Contrast as a function of the parameters a and d. The units on the axes a and d are respectively 0.22 and 0.15.

This figure yields the following conclusions

**a.** As expected the contrast decreases while a increases for any d. When a = 0 the receiver becomes linear then we see that the contrast is always below this of the linear receiver. The degradation of the contrast may become very important and it is seen that when a>3, the contrast reaches almost zero whatever d. The saturation effect of the nonlinear transformation suppresses completely any contrast.

**b.** For a = 0, the contrast curve is a parabola, which is a direct consequence of the fact that in this case the contrast is the signal to noise ratio $d^2$.

c. The study of the asymptotic behavior of the contrast confirms the saturation effect due to the nonlinearity. As a matter of fact, it is possible to deduce from (5-8) that

$$\lim_{d \to +\infty} C[d,a] = C[\infty,a] = k^2 \tag{5-9}$$

where k, function of a, is the normalization constant given by (5-6). As a is decreasing to zero $C[\infty,a]$ becomes infinite, which is satisfactory, because $S_a[x]$ becomes the linear receiver. On the other hand, as soon as a is not equal to zero, $C[\infty,a]$ keeps finite values, which shows clearly the degradation due to the nonlinearity.

### 5.2. **Modification of the detection performance**

Our purpose is to compare the behaviour of the contrast after a monotonic transformation and the detection performance after quantization. This performance is characterized by the ROC curves defined at the end of section 2.

In this section the output of the monotonic transformation $S_a[x]$ defined by (5-5) and (5-6) is quantized, as indicated in 5.1, and then the detection probability is

$$\beta = f(\alpha;\ a,\ d,\ q)\ , \tag{5-10}$$

which means that it is a function of α but also of the parameters of the problem : a, defining the monotonic transformation, d, defining the signal to noise ratio and q, defining the quantization step. As q = 0, which means that there is no quantization, we get the classical R.O.C. curves of the matched filter, (see [14], p.107 or [25] p.38). Instead of drawing directly β, it is better to represent Δβ, difference between detection probabilities of the linear receiver and the studied receiver. As the linear receiver is reached when a = 0, we have of course

$$\Delta\beta = f[\alpha;\ o,\ d,\ q] - f[\alpha;\ a,\ d,\ q]. \tag{5-11}$$

All the numerical calculations have shown that Δβ is positive, which means that the linear receiver is always a better detector, a point already noticed on the contrast curves.

The quantity Δβ may represent the degradation of the performance due to the monotonic transformation in presence of quantization, which can depend upon 4 free parameters $\alpha$, a, d, q. For a three dimensional display two of these parameters are kept constant.

In figure 4 are presented the variations of $\Delta\beta$ as a function of a, d for a given value of $\alpha$ and q. On the other hand $\alpha$ and a are given in figure 5.

These two figures, similar to many others giving the same behaviour and studied elsewhere, lead to the following conclusions.

**a.** Whatever d, $\Delta\beta$ increases with a or q. The increase with a is in relation to the decrease of the contrast noticed in figure 3. Consequently, for a given quantization step, a better contrast corresponds to an inprovement of detection performance. Moreover for a given value of a, i.e. for a given contrast, the performance is always decreasing as q is increasing.

**b.** We notice from figure 4 that within the range of the nonlinearity studied here, $\Delta\beta$ is reaching zero as d = 0. This fact means that all the receivers are equivalent when d is equal to zero. On the other hand when d becomes very large, we would expect the same behaviour. A cautious analysis of this situation shows that it is not true. Indeed when $a \to \infty$, which means that we use the sign receiver, the difference $\Delta\beta$, called here $\Delta\beta_\infty$, satisfies

$$\lim_{d \to +\infty} \Delta\beta_\infty \geq 1-2\,\alpha\,, \qquad (5\text{-}12)$$

where $\alpha$ is the false alarm probability. This relation shows how the saturation effect, due to the nonlinearity, damages the detection performances : even if the signal to noise ratio increases, the nonlinearity suppresses any improvement of the performance. This phenomenon has already been indicated on the contrast variations (see commentary c on figure 3).

Finally we observe in figures 4 and 5 that the behaviours of $\Delta\beta$ for a given value of d are similar if a or q is increasing. But as a is directly connected with the contrast, as seen in figure 3, we can say that in our example a better contrast gives better performance in quantized detection. Of course without quantization the performance is independant of the contrast.

This gives a new connection between statistical detection theory and the contrast theory. Of course this property appears only on the particular example presented here. It would be very interesting to check if that is a general property, but the problem appears extremely difficult.

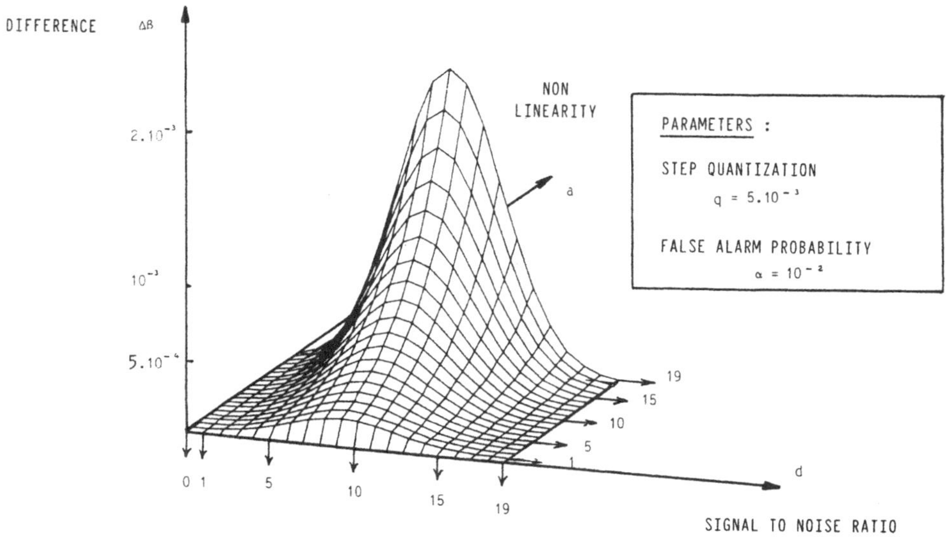


Figure 4 : Difference Δβ for a given step quantization and false alarm probability.

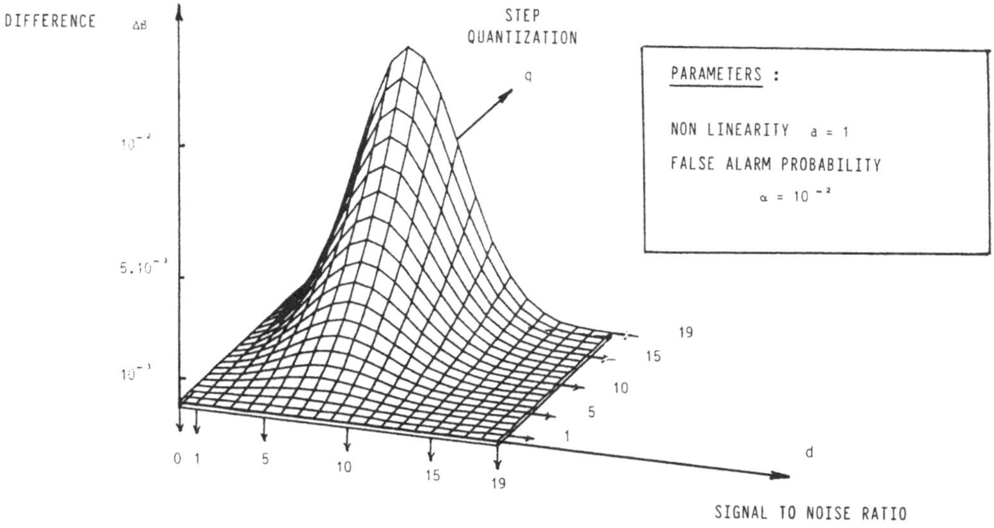


Figure 5 : Difference Δβ for a given value of the nonlinearity and false alarm probability.

## REFERENCES


[1] E.K. AL-HUSSAINI and S.A. KASSAM, "Robust Eckart filter for time delay estimation," IEEE Trans. Acoust. Speech, Signal Processing, ASSP-29, pp.1052-1063, October 1984.

[2] C.R. BAKER, "Optimum quadratic detection of a random vector in Gaussian noise," IEEE Trans. Comm., COM-14, pp.802-805, Dec.1966.

[3] M. BELLANGER, Traitement numérique du signal, Masson, Paris, 1984.

[4] A. BLANC-LAPIERRE et B. PICINBONO, Propriétés statistiques du bruit de fond, Paris, Masson, 1960, p.81.

[5] J. CAPON, "On the asymptotic efficiency of locally optimum detectors," IRE Trans. Inf. Theory, IT 7, pp.67-71, April 1961.

[6] S.V. CZARNECKI and J. THOMAS, "Nearly optimal detection of signals in non-Gaussian noise," Rep. N° 14, Information science and system laboratory, Princeton University, Feb. 1984.

[7] S.V. CZARNECKI and K.S. VASTOLA, "Approximation of locally optimum detectors non linearities," 1983 Conference on information sciences and systems, Johns Hopkins University, Baltimore, Maryland, March 1983.

[8] C. ECKART, "Optimal rectifier systems for the detection of steady signals," Techn. Rep. S 10, Ref. 52-11, Scripps Institue of Oceanography, University of California, March 1952.

[9] H. EL AYADI and B. PICINBONO, "QNAR AGC system for adaptive signal detection," IEEE Trans. ASSP, ASSP-31, pp.225-228, Feb. 1983.

[10] H. EL AYADI and B. PICINBONO, "NAR AGC adaptive detection of nonoverlapping signals in noise with fluctuating power," IEEE Trans. ASSP, ASSP-29, pp.952-963, Oct. 1981.

[11] W.A. GARDNER, "Anomalous behavior of receiver output SNR as a predictor of signal detection performance exemplified for quadratic receivers and incoherent fading Gaussian channel," IEEE Trans. Inf. Theory, IT 25, pp.743-745, Nov. 1979.

[12] W.A. GARDNER, "A unifying view of second-order measures of quality for signal classification," IEEE Trans. Comm. COM-28, pp.807-816, June 1980.

[13] J. GIBSON and J. MELSA, Introduction to nonparametric detection with application, Academic Press, New York, 1975.

[14] C.W. HELSTROM, Statistical theory of signal detection, Oxford, Pergamon, 1968.

[15] J.L. LAWSON and G.E. UHLENBECK, Threshold signals, New-York, McGraw Hill, 1950.

[16] V.H. MAC DONALD and P.P. SCHULTHEISS, "Optimum passive bearing estimation," J. Acoust. Soc. Amer. 46, pp.37-43, July 1969.

[17] G. MEYER and H. WEINERT, "The effects of hardware faults on signal detector performance," 23° IEEE Conference on Decision and Control, Las Vegas, 1984, p.642-643.

[18] B. PICINBONO, "A geometrical interpretation of signal detection and estimation," IEEE Trans. Inf. Theory, IT 26, pp.493-497, July 1980.

[19] B. PICINBONO and P. DUVAUT, "Nouvelle approche de la détection par seuil," 9ème Colloque GRETSI, Nice, 1983, pp.87-91.

[20] E.J. PITMAN, Some basic theory of statistical inference, London, Chapman and Hall, 1979.

[21] H.V. POOR, "Robust decision design using a distance criterion," IEEE Trans. Inf. Theor. IT 26, pp.575-587, Sept. 1980.

[22] P. RUDNICK, "A signal to noise property of binary decision," Nature, Vol. 193, pp.604-605, Fev. 1962.

[23] P. RUDNICK, "Likelihood detection of small signals in stationary noise," J. Appl. Physics, 32, pp.140-143, Feb. 1961.

[24] J. THOMAS, "Nonparametric detection," Proc. of the IEEE, 58, pp.623-631, 1970.

[25] H.L. VAN TREES, Detection, estimation and modulation theory, Part. 1, New-York, Wiley, 1968.

[26] S. VERDU, "Comment on Ref.[18]," IEEE Trans. Inf. Theory, IT-28, pp.952-953, Nov. 1982.

## SUBJECT INDEX

# Lecture Notes in Control and Information Sciences

Edited by M. Thoma

Vol. 43: Stochastic Differential Systems
Proceedings of the 2nd Bad Honnef Conference
of the SFB 72 of the DFG at the University of Bonn
June 28 – July 2, 1982
Edited by M. Kohlmann and N. Christopeit
XII, 377 pages. 1982.

Vol. 44: Analysis and Optimization of Systems
Proceedings of the Fifth International
Conference on Analysis and Optimization of Systems
Versailles. December 14–17, 1982
Edited by A. Bensoussan and J. L. Lions
XV, 987 pages, 1982

Vol. 45: M. Arató
Linear Stochastic Systems
with Constant Coefficients
A Statistical Approach
IX, 309 pages. 1982

Vol. 46: Time-Scale Modeling of Dynamic Networks
with Applications to Power Systems
Edited by J. H. Chow
X, 218 pages. 1982

Vol. 47: P. A. Ioannou, P. V. Kokotovic
Adaptive Systems with Reduced Models
V, 162 pages. 1983

Vol. 48: Yaakov Yavin
Feedback Strategies for Partially
Observable Stochastic Systems
VI, 233 pages, 1983

Vol. 49: Theory and Application of Random Fields
Proceedings of the IFIP-WG 7/1
Working Conference
held under the joint auspices of the
Indian Statistical Institute
Bangalore, India, January 1982
Edited by G. Kallianpur
VI. 290 pages. 1983

Vol. 50: M. Papageorgiou
Applications of Automatic Control Concepts
to Traffic Flow Modeling and Control
IX, 186 pages. 1983

Vol. 51: Z. Nahorski, H.F. Ravn, R.V.V. Vidal
Optimization of Discrete Time Systems
The Upper Boundary Approach
V, 137 pages 1983

Vol. 52: A. L. Dontchev
Perturbations, Approximations and Sensitivity Analysis
of Optimal Control Systems
IV, 158 pages. 1983

Vol. 53: Liu Chen Hui
General Decoupling Theory of Multivariable
Process Control Systems
XI, 474 pages. 1983

Vol. 54: Control Theory for Distributed
Parameter Systems and Applications
Edited by F. Kappel, K. Kunisch,
W. Schappacher
VII, 245 pages. 1983.

Vol. 55: Ganti Prasada Rao
Piecewise Constant Orthogonal Functions
and Their Application to Systems and Control
VII, 254 pages. 1983.

Vol. 56: Dines Chandra Saha, Ganti Prasada Rao
Identification of Continuous
Dynamical Systems
The Poisson Moment Functional
(PMF) Approach
IX, 158 pages. 1983.

Vol. 57: T. Söderström, P. G. Stoica
Instrumental Variable Methods
for System Identification
VII, 243 pages. 1983.

Vol. 58: Mathematical Theory of
Networks and Systems
Proceedings of the MTNS-83 International
Symposium
Beer Sheva, Israel, June 20–24, 1983
Edited by P. A. Fuhrmann
X, 906 pages. 1984

Vol. 59: System Modelling and Optimization
Proceedings of the 11th IFIP Conference
Copenhagen, Denmark, July 25-29, 1983
Edited by P. Thoft-Christensen
IX, 892 pages. 1984

Vol. 60: Modelling and Performance
Evaluation Methodology
Proceedings of the International Seminar
Paris, France, January 24–26, 1983
Edited by F. Bacelli and G. Fayolle
VII, 655 pages. 1984

Vol. 61: Filtering and Control of Random
Processes
Proceedings of the E.N.S.T.-C.N.E.T. Colloquium
Paris, France, February 23–24, 1983
Edited by H. Korezlioglu, G. Mazziotto, and
J. Szpirglas
V, 325 pages. 1984

# Lecture Notes in Control and Information Sciences

Edited by M. Thoma and A. Wyner

Vol. 62: Analysis and Optimization of Systems
Proceedings of the Sixth International Conference on Analysis and Optimization of Systems
Nice, June 19–22, 1984
Edited by A. Bensoussan, J. L. Lions
XIX, 591 pages. 1984.

Vol. 63: Analysis and Optimization of Systems
Proceedings of the Sixth International Conference on Analysis and Optimization of Systems
Nice, June 19–22, 1984
Edited by A. Bensoussan, J. L. Lions
XIX, 700 pages. 1984.

Vol. 64: Arunabha Bagchi
Stackelberg Differential Games in Economic Models
VIII, 203 pages, 1984

Vol. 65: Yaakov Yavin
Numerical Studies in Nonlinear Filtering
VIII, 273 pages, 1985.

Vol. 66: Systems and Optimization
Proceedings of the Twente Workshop
Enschede, The Netherlands, April 16–18, 1984
Edited by A. Bagchi, H. Th. Jongen
X, 206 pages, 1985.

Vol. 67: Real Time Control of Large Scale Systems
Proceedings of the First European Workshop
University of Patras, Greece, Juli 9–12, 1984
Edited by G. Schmidt, M. Singh, A. Titli, S. Tzafestas
XI, 650 pages, 1985.

Vol. 68: T. Kaczorek
Two-Dimensional Linear Systems
IX, 397 pages, 1985.

Vol. 69: Stochastic Differential Systems – Filtering and Control
Proceedings of the IFIP-WG 7/1 Working Conference
Marseille-Luminy, France, March 12-17, 1984
Edited by M. Metivier, E. Pardoux
X, 310 pages, 1985.

Vol. 70: Uncertainty and Control
Proceedings of a DFVLR International Colloquium
Bonn, Germany, March, 1985
Edited by J. Ackermann
IV, 236 pages, 1985.

Vol. 71: N. Baba
New Topics in Learning Automata Theory and Applications
VII, 231 pages, 1985.

Vol. 72: A. Isidori
Nonlinear Control Systems: An Introduction
VI, 297 pages, 1985.

Vol. 73: J. Zarzycki
Nonlinear Prediction Ladder-Filters for Higher-Order Stochastic Sequences
V, 132 pages, 1985.

Vol. 74: K. Ichikawa
Control System Design based on Exact Model Matching Techniques
VII, 129 pages, 1985.

Vol. 75: Distributed Parameter Systems
Proceedings of the 2nd International Conference, Vorau, Austria 1984
Edited by F. Kappel, K. Kunisch, W. Schappacher
VIII, 460 pages, 1985.

Vol. 76: Stochastic Programming
Edited by F. Archetti, G. Di Pillo, M. Lucertini
V, 285 pages, 1986.

Vol. 77: Detection of Abrupt Changes in Signals and Dynamical Systems
Edited by M. Basseville, A. Benveniste
X, 373 pages, 1986.

Vol. 78: Stochastic Differential Systems
Proceedings of the 3rd Bad Honnef Conference, June 3–7, 1985
Edited by N. Christopeit, K. Helmes, M. Kohlmann
V, 372 pages, 1986.

Vol. 79: Signal Processing for Control
Edited by K. Godfrey, P. Jones
XVIII, 413 pages, 1986.

Vol. 80: Artificial Intelligence and Man-Machine Systems
Edited by H. Winter
IV, 211 pages, 1986.

# Lecture Notes in Control and Information Sciences

Edited by M. Thoma and A. Wyner

Vol. 81: Stochastic Optimization
Proceedings of the International Conference,
Kiew, 1984
Edited by I. Arkin, A. Shiraev, R. Wets
X, 754 pages, 1986.

Vol. 82: Analysis and Algorithms
of Optimization Problems
Edited by K. Malanowski, K. Mizukami
VIII, 240 pages, 1986.

Vol. 83: Analysis and Optimization
of Systems
Proceedings of the Seventh International
Conference of Analysis and Optimization
of Systems
Antiba, June 26-27, 1986
Edited by A. Benoussan, J. L. Lions
XVI, 901 pages, 1986.

Vol. 84: System Modelling
and Optimization
Proceedings of the 12th IFIP Conference
Budapest, Hungary, September 2–6, 1985
Edited by A. Prékopa, J. Szelezsán, B. Strazicky
XII, 1046 pages, 1986.

Vol. 85: Stochastic Processes
in Underwater Acoustics
Edited by C. R. Baker
V, 205 pages, 1986.